NUMERICAL ADVENTURES
WITH GEOCHEMICAL CYCLES

NUMERICAL ADVENTURES
WITH GEOCHEMICAL CYCLES

James C. G. Walker

New York   Oxford
OXFORD UNIVERSITY PRESS
1991

Oxford University Press

Oxford   New York   Toronto
Delhi   Bombay   Calcutta   Madras   Karachi
Petaling Jaya   Singapore   Hong Kong   Tokyo
Nairobi   Dar es Salaam   Cape Town
Melbourne   Auckland

and associated companies in
Berlin   Ibadan

Published by Oxford University Press, Inc.,
200 Madison Avenue, New York, New York 10016

Oxford is a registered trademark of Oxford University Press

Library of Congress Cataolging-in-Publication Data
Walker, James C. G. (James Callan Gray)
Numerical adventures with geochemical cycles/James C. G. Walker.
p. cm.   Includes bibliographical references and index.
ISBN 0-19-504520-3
1. Analytic geochemistry. 2. Geology—Periodicity.
3. Chemical oceanography. 4. Numerical calculations.
I. Title.   QE516.3.W35 1991   551.9--dc20   90-39135

9 8 7 6 5 4 3 2 1

Printed in the United States of America
on acid-free paper

Acknowledgments

I am most grateful to the following individuals who have contributed either in person or through their research to the development of the ideas I present here: Wally Broecker, Bruce Wilkinson, K. C. Lohmann, and Brad Opdyke. Jim Kasting taught me how to solve the kind of differential equation that arises in the simulation of global change; without his help I could not have written this book. Tom Algeo provided detailed and helpful criticism of an earlier draft of this book, and his advice led to its substantial improvement. The computational methods were tested by several groups of students over a period of years at the University of Michigan. I learned much from them.

Contents

NUMERICAL ADVENTURES WITH GEOCHEMICAL CYCLES

1 Why Simulate?

Our world is a product of complex interactions among atmosphere, ocean, rocks, and life that Earth system science seeks to understand. Earth system science deals with such properties of the environment as composition and climate and populations and the ways in which they affect one another. It also concerns how these interactions caused environmental properties to change in the past and how they may change in the future. The Earth system can be studied quantitatively because its properties can be represented by numbers. At present, however, most of the numbers in Earth system science are observational rather than theoretical, and so our description of the Earth system's objective properties is much more complete than our quantitative understanding of how the system works. Quantitative theoretical understanding grows out of a simulation of the system or parts of the system and numerical experimentation with simulated systems. Simulation experiments can answer questions like What is the effect of this feature? or What would happen in that situation? Simulation also gives meaning to observations by showing how they may be related. As an illustration, consider that area of Earth system science known as *global change*.

There is now an unambiguous observational record of global change in many important areas of the environment. For elements of climate and atmospheric composition this record is based on direct measurement over periods of a decade to a century. For other environmental variables, particularly those related to the composition of the ocean, the record of change consists of measurements of isotopic or trace-element composition of sediments deposited over millions of years. This evidence of global change is profoundly affecting our view of what the future holds in store for us and what options exist. It should also influence our understanding of how the interaction of biota and environment has changed the course of Earth history. But despite the importance of global change to our prospects for the future and our understanding of the past, the mechanisms of change are little understood. There are many speculative suggestions about the causes of change but few quantitative and convincing tests of these suggestions.

Theoretical simulations of environmental systems can provide such tests, by demonstrating in quantitative terms the consequences, according to our current understanding, of various postulated changes. The use of simulation to study the climate system is well known. Climate models are now exceedingly complex and sophisticated and are routinely used for studying climatic change associated with human activities, ice ages, or geological evolution, as well as for predicting the weather. Simulations of the minor constituent composition of the atmosphere have also been developed to considerable degrees of complexity. These are the simulations used, for example, to study the possible effects of human activities on atmospheric ozone. Atmospheric photochemical models are also used in studies of acid rain and the compositions of other planets' atmospheres.

Simulations of the major constituent composition of the atmosphere or of the composition of the ocean are much less highly developed. The reason is that the time scales for change in the composition of the ocean and the major constituents of the atmosphere are long in human terms, and therefore it is more difficult to observe the processes and mechanisms of change. There also is less need to understand the causes and consequences of change. As a result, the evolution of ocean and atmosphere has as yet received little quantitative theoretical attention. Fundamental problems in global change are thus waiting for original research using computational simulations of the composition of ocean and atmosphere, and this book shows how to develop such simulations. It presents computational methods that can be applied not only to global change in the composition of the ocean and atmosphere but also to climate change and properties of more restricted terrestrial environments such as lagoons or sediments.

The book explains how to solve coupled systems of ordinary differential equations of the kind that commonly arise in the quantitative description of the evolution of environmental properties. All of the computations that I shall describe can be performed on a personal computer, and all of the programs can be written in such familiar languages as BASIC, PASCAL, or FORTRAN. My goal is to teach the methods of computational simulation of environmental change, and so I do not favor the use of professionally developed black-box programs.

It turns out to be quite easy to write computer programs that can be applied to many aspects of Earth system evolution, that are robust and well behaved, and that are friendly—indeed transparent—to the user. My approach emphasizes methods that are easy to understand and to use, rather than methods of great sophistication and precision. The user should spend time investigating the system, not the methods of solution.

I apply these computational methods to various aspects of the Earth system, including the responses of ocean and atmosphere to the combustion of fossil fuels, the influence of biological activity on the variation of seawater composition between ocean basins, the oxidation–reduction balance of the deep sea, perturbations of the climate system and their effect on surface temperatures, carbon isotopes and the influence of fossil fuel combustion, the effect of evaporation on the composition of seawater, and diagenesis in carbonate sediments. These applications have not been fully developed as research studies; rather, they are presented as potentially interesting applications of the computational methods.

Chapter 2 shows why the most obvious approach to calculating the evolution of a quantity like atmospheric carbon dioxide is not generally useful because of numerical instability. The chapter introduces instead a numerical technique that is robust and flexible and provides the basis for all of the work that follows. Chapter 3 deals with coupled systems, beginning with a coupled system of linear algebraic equations that describe the variation of ocean composition between basins in the steady state. This chapter shows how the solution of the algebraic equations can be used to solve the system of linear ordinary differential equations that describes the evolution of the same system when it is not in a steady state. Chapter 4 further explains the computational method by showing how it may be used for a nonlinear system of coupled ordinary differential equations. The application is to the oxidation–reduction balance of the deep sea. The method is to approximate the nonlinear system by the equivalent linear system and to progress in time steps that are small enough to make the linear approximation accurate. The chapter presents an easy method of testing and adjusting the time step in order to ensure that this condition is met. The chapter also shows how to automate the linearization procedure so that the researcher can concentrate on formulating and understanding the system without having to worry unduly about algebraic manipulation.

Chapter 5 introduces the equations that describe equilibrium between dissolved species of carbon in the ocean. The carbon system is central to so many aspects of ocean, atmosphere, and sedimentary rocks that I wish to make available a general method of solution. This chapter also introduces several housekeeping routines that I have found useful. Examples are routines to read files of specifications, including initial values, to write files of results for subsequent plotting, and to provide a flexible graphical presentation of results on the computer screen during a calculation. As an application, I consider the response of carbon in ocean and atmosphere to the combustion of fossil fuels. Chapter 6 takes up the calculation of isotope ratios. Isotopes provide one of the most useful sources of information about the history of ocean and atmosphere and about the relative importance of various processes. An isotope calculation requires special consideration because isotopic composition is generally described by ratios of rare isotope to abundant isotope, and ratios are not conserved in quite the same way as amounts are. In this chapter, I explain how equations for isotope ratios can be derived from the equations for the corresponding concentrations, provided that a few extra terms are included. As applications I consider the effect of fossil fuel combustion on the isotopic composition of ocean and atmosphere and also simulate a hypothetical lagoon from which water is evaporating and calcium carbonate is precipitating. The simulation demonstrates how the composition of the water in the lagoon is affected by evaporation, precipitation, and the exchange of carbon with the atmosphere.

Significant economies of computation are possible in systems that consist of a one-dimensional chain of identical reservoirs. Chapter 7 describes such a system in which there is just one dependent variable. An illustrative example is the climate system and the calculation of zonally averaged temperature as a function of latitude in an energy balance climate model. In such a model, the surface temperature depends on the balance among solar radiation absorbed, planetary radiation emitted to space, and the transport of energy between latitudes. I present routines that calculate the absorption and reflection of incident solar radiation and the emission of long-wave planetary radiation. I show how much of the computational work can be avoided in a system like this because each reservoir is coupled only to its adjacent reservoirs. I use the simulation to explore the sensitivity of seasonally varying temperatures to such aspects of the climate system as snow and ice cover, cloud cover, amount of carbon dioxide in the atmosphere, and land distribution.

Chapter 8 describes a similar one-dimensional chain of identical reservoirs, but one that contains several interacting species. The example illustrated here is the composition of the pore waters in carbonate sediments in which dissolution is occurring as a result of the oxidation of organic matter. I calculate the concentrations of total dissolved carbon and calcium ions and the isotope ratio as functions of depth in the sediments. I present modified routines for the efficient solution of systems consisting of interacting species in a chain of identical reservoirs.

I want my readers to try out my computational methods. All of the work that I describe in this book I did on an IBM PS/2 Model 50Z computer with a math coprocessor, and all of my calculations were completed in less than half an hour. If your machine is much slower than mine, you should expect to wait for the results from some of the longer calculations presented in the later chapters. I use a compiled BASIC because the BASIC interpreter that comes with the computer is too slow for all but the shortest calculations. Most of the programs given in this book are modifications of programs introduced earlier. The changes in each program are indicated by statements in bold italic type. When subroutines are carried forward without change I do not list them again but simply refer back to the program in which they first appeared. In the text I use italic lowercase letters for program variables and Roman capital letters for the names of programs, files, and subroutines.

2 How to Calculate a Compositional History

2.1 Introduction

The most interesting theoretical problems in Earth system science cannot be solved by analytical methods; their solutions cannot be expressed as algebraic expressions; and so numerical solutions are needed. In this chapter I shall introduce a method of numerical solution that can be applied to a wide range of simulations and yet is easy to use. In later chapters I shall elaborate and apply this method to a variety of situations.

All numerical solutions of differential equations involve some degree of approximation. Derivatives—properly defined in terms of infinitesimally small increments—are approximated by finite differences: dy/dx is approximated by *dely/delx,* whose accuracy increases as the finite difference, *delx,* is reduced. A large value of *delx* may even cause numerical instability, yielding a numerical solution that is altogether different from the true solution of the original system of differential equations. But a small value of *delx* generally implies that many steps must be taken to evolve the solution from a starting value of x to a finishing value. A numerical solution, therefore, requires a trade-off between computational speed and accuracy. We seek an efficient and stable method of calculation that provides accuracy appropriate to our knowledge of the physical system being simulated. The problem described in this chapter can be easily solved analytically, and the analytical solution serves as a check on the accuracy of the numerical solutions.

2.2 A Simple Simulation

As a simple example of a global geochemical simulation, consider the exchange of carbon dioxide between the ocean and the atmosphere. The atmosphere contains 5.6×10^{16} moles of carbon dioxide (cf. Walker, 1977), a quantity that I assume to be in equilibrium with the ocean. In this illustration I assume that the oceanic reservoir is very large and therefore does not change with time. According to Broecker and Peng (1982, p. 680) the annual exchange of carbon dioxide between ocean and atmosphere is 6.5×10^{15} moles. The rate of transfer from atmosphere to ocean is proportional to the amount in the atmosphere; the flow from ocean to atmosphere is constant. Figure 2–1 summarizes this system.

The amount of carbon dioxide in the atmosphere is proportional to the partial pressure. Thus if partial pressure is expressed in units of the present level, the amount will become $56 \times 10^{15} * pco2$ moles, where *pco2* is the partial pressure of carbon dioxide. Expressing the reservoir in units of 10^{15} moles and the exchange rate in 10^{15} moles/y, the equation for the time rate of change of carbon dioxide partial pressure, assuming no other sources or sinks, becomes

$$56 \, \frac{d \, pco2}{d \, t} = 6.5 * (1 - pco2)$$

or

$$\frac{d \, pco2}{d \, t} = (1 - pco2)/distime$$

where

$$distime = 56/6.5 = 8.64 \text{ years}$$

is the average length of time that carbon dioxide remains in the atmosphere before dissolving in the ocean. This parameter is called the *residence time,* and it is equal to the reservoir amount divided by the source rate.

The system is taken to be in a steady state at present, so $d(pco2)/dt = 0$ when $pco2 = 1$. I shall next examine the recovery from an assumed sudden injection of a large amount of fossil fuel carbon. According to Broecker and Peng (1982) the total fossil fuel release of carbon dioxide over the next 400 years might amount to 2.3×10^{17} moles, approximately four times the amount now in the atmosphere. If all of this carbon dioxide were released at once, without time for its transfer to the ocean, the partial pressure of atmospheric carbon dioxide would rise to five times the present level. I shall therefore simulate this situation, the recovery of *pco2* from a starting value of 5, an exercise that illustrates computational principles, not the real world. A more realistic examination of the response of atmosphere and ocean to fossil fuel carbon dioxide will appear in Chapter 5.

This system can, of course, be solved analytically. The solution is

$$pco2 = 1 + 4 * \exp(-t/distime)$$

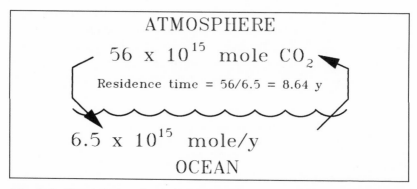

Fig. 2–1. The exchange of carbon dioxide between the atmosphere and an infinite oceanic reservoir.

where t is the time after the sudden injection of carbon dioxide. More complicated simulations will not be susceptible to analytical solution, so I shall seek a general numerical approach. The analytical solution to this particularly simple simulation serves as a check on the numerical solution.

2.3 The Obvious Approach

For generality, let $y = pco2$ be the partial pressure of carbon dioxide and let $x = t$ be the time. The equation is

$$\frac{dy}{dx} = yp(x,y)$$

where

$$yp(x,y) = (1 - y(x))/distime$$

The value of y is known at some time x, and its value is sought at some future time $x + delx$. The obvious approach, which is called the *direct Euler method,* is to approximate the function by a straight line of slope $yp(x,y(x))$ and to calculate the new value of y from

$$y(x + delx) = y(x) + delx * yp(x,y(x))$$

The calculation is repeated to step forward in time.

Program DGC01 implements this method of solution. Although the use of subroutines may appear unnecessarily elaborate in such a small program, I am introducing them now to facilitate future development of the program.

```
'Program DGC01 is direct Euler solution of carbon dioxide exchange
'between atmosphere and ocean, applied to sudden fossil fuel input
x = 0          'starting time
delx = 20         'time step
nstep = 10        'number of steps
matmco2 = .056
distime = 8.64
pco2 = 5          'starting value
y = pco2
OPEN "RESULTS.PRN" FOR OUTPUT AS #1    'results to a file for plotting
PRINT "YEARS", "PCO2"
PRINT x, y
PRINT #1, x, y
FOR nx = 1 TO nstep
    GOSUB EQUATIONS
    GOSUB STEPPER
    GOSUB PRINTER
NEXT nx
CLOSE #1
END
'*********************************************************************
PRINTER:    'Subroutine writes to a file for subsequent plotting
pco2 = y
PRINT x, pco2
PRINT #1, x, pco2
'*********************************************************************
EQUATIONS: 'Subroutine to calculate the slope, yp
pco2 = y
yp = (1 - pco2) / distime
RETURN
'*********************************************************************
STEPPER:    'Subroutine steps forward in time
y = y + yp * delx
x = x + delx
RETURN
```

Figure 2–2 shows the results calculated with this program. The heavy solid line is the analytical solution, and the lines with markers on them show the results for *delx* = 3 and 10 years. The inset shows, over a longer time span, the behavior of the numerical solution for *delx* = 20 years. This

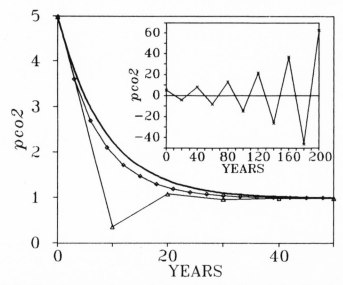

Fig. 2–2. The recovery of atmospheric carbon dioxide calculated by the direct Euler method. The solid line is the analytical solution, and the lines with markers show the numerical results calculated with program DGC01 using time steps *delx* = 3, 10, and 20 years. The numerical solution is unstable for time steps of 10 years or longer.

computational method works well for time steps smaller than the 8.6-year residence time but displays strong instability for larger time steps. The method thus would not be useful for calculating the history of carbon dioxide over geological time spans of hundreds of millions of years, as too many time steps would be needed.

2.4 A Scheme That Works

This instability can be eliminated by calculating the slope of the straight-line approximation not at *x*, where the value of *y* is already known, but at the future time *x* + *delx*. Therefore, instead of

$$y(x + delx) = y(x) + delx * yp(x,y(x))$$

use

$$y(x + delx) = y(x) + delx * yp(x + delx,y(x + delx))$$

```basic
'Program DGCO2 is reverse Euler solution of carbon dioxide exchange
'between atmosphere and ocean, applied to sudden fossil fuel input
x = 0
delx = 20
nstep = 10
matmco2 = .056
distime = 8.64
pco2 = 5          'starting value
y = pco2
OPEN "RESULTS.PRN" FOR OUTPUT AS #1     'results to a file for plotting
PRINT "YEARS", "PCO2"
PRINT x, y
PRINT #1, x, y
FOR nx = 1 TO nstep
    GOSUB EQUATIONS
    GOSUB STEPPER
    GOSUB PRINTER
NEXT nx
CLOSE #1
END
'**************************************************************************
PRINTER:    'Subroutine writes a file for subsequent plotting
pco2 = y
PRINT x, pco2
PRINT #1, x, pco2
RETURN
'**************************************************************************
EQUATIONS: 'Subroutine to calculate the slope, yp
pco2 = y
yp = (1 - pco2) / (distime + delx)
RETURN
'**************************************************************************
STEPPER:    'Subroutine steps forward in time
y = y + yp * delx
x = x + delx
RETURN
```

where, with $yp = (1 - y)/distime$,

$yp(x + delx, y(x + delx))$
$$= (1 - (y(x) + delx * yp(x + delx, y(x + delx))))/distime$$

Solve for $yp(x + delx, y(x + delx))$ to obtain

$$yp(x + delx, y(x + delx)) = (1 - y(x))/(distime + delx)$$

Program DGC02 implements this solution, which is called the *reverse Euler method*. The only change in the code is in the expression for yp in the subroutine EQUATIONS.

The results are shown in Figure 2–3, in which the solid line is the exact solution. This numerical approach shows no sign of instability even for a time step of 40 years, nearly five times larger than the residence time of atmospheric carbon dioxide (*distime*). In fact, the reverse Euler method is nearly always stable, and so I shall use it from now on.

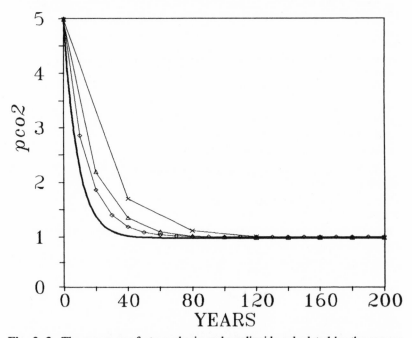

Fig. 2–3. The recovery of atmospheric carbon dioxide calculated by the reverse Euler method. The solid line is the analytical solution, and the lines with markers show the numerical results calculated with program DGC02 using time steps *delx* = 10, 20, and 40 years. The numerical solution is stable for all time steps.

2.5 Summary

A high degree of accuracy is not called for in many calculations of the evolution of environmental properties because the mathematical description of the environment by a reasonably small number of equations involves an approximation quite independent of any approximation in the equations' solution. Figure 2–3 shows how the accuracy of the reverse Euler method degrades as the time step is increased, but it also shows the stability of the method. Even a time step of 40 years, nearly five times larger than the residence time of 8.64 years, yields a solution that behaves like the true solution. In contrast, Figure 2–2 shows the instability of the direct Euler method; a time step as small as 10 years introduces oscillations that are not a property of the true solution.

More complicated numerical methods, such as the Runge–Kutta method, yield more accurate solutions, and for precisely formulated problems requiring accurate solutions these methods are helpful. Examples of such problems are the evolution of planetary orbits or the propagation of seismic waves. But the more accurate numerical methods are much harder to understand and to implement than is the reverse Euler method. In the following chapters, therefore, I shall show the wide range of interesting environmental simulations that are possible with simple numerical methods.

The key feature of the systems to be considered in this book is that they have short memories; that is, the effects of perturbations diminish with the passage of time. In the example of this chapter, the carbon dioxide pressure returns to a value of 1 within a century or two of the perturbation, regardless of the size of the initial perturbation. In this kind of system, computational errors do not grow as the calculation proceeds; instead, the system forgets old errors. That is why the reverse Euler method is useful despite its simplicity and limited accuracy. The many properties of the environment that are reasonably stable and predictable can, in principle, be described by equations with just this kind of stability, and these are the properties that can be simulated using the computational methods described in this book.

3 How to Deal with Several Reservoirs

3.1 Introduction

The previous chapter showed how the reverse Euler method can be used to solve numerically an ordinary first-order linear differential equation. Most problems in geochemical dynamics involve systems of coupled equations describing related properties of the environment in a number of different reservoirs. In this chapter I shall show how such coupled systems may be treated. I consider first a steady-state situation that yields a system of coupled linear algebraic equations. Such a system can readily be solved by a method called *Gaussian elimination* and *back substitution*. I shall present a subroutine, GAUSS, that implements this method.

The more interesting problems tend to be neither steady state nor linear, and the reverse Euler method can be applied to coupled systems of ordinary differential equations. As it happens, the application requires solving a system of linear algebraic equations, and so subroutine GAUSS can be put to work at once to solve a linear system that evolves in time. The solution of nonlinear systems will be taken up in the next chapter.

3.2 A System of Coupled Reservoirs

Most simulations of environmental change involve several interacting reservoirs. In this chapter I shall explain how to apply the numerical scheme described in the previous chapter to a system of coupled equations. Figure 3–1, adapted from Broecker and Peng (1982, p. 382), is an example of a coupled system. The figure presents a simple description of the general circulation of the ocean, showing the exchange of water in Sverdrups (1 Sverdrup = 10^6 m^3/sec) among five oceanic reservoirs and also the addi-

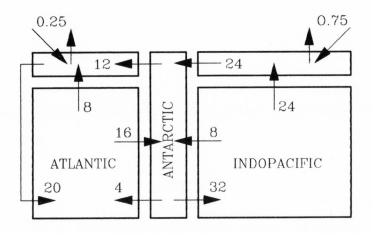

Fig. 3–1. A simple representation of the general circulation of the global ocean, adapted from Broecker and Peng (1982, p. 382). The arrows denote fluxes of water expressed in Sverdrups (1 Sverdrup = 10^6 m³/sec). River water is added to the surface reservoirs, and an equal volume of water is removed by evaporation.

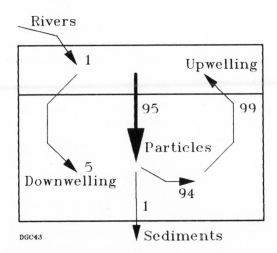

Fig. 3–2. I assume that 95 percent of the phosphorus supplied to the surface sea is incorporated into organic matter and returned to the deep sea in particulate form. One percent of the total survives to be buried in sediments. The rest is restored to the deep sea as dissolved phosphorus. The loss to sediments is balanced for the whole ocean by supply by the rivers. The fluxes here are in relative units.

17

tion of river water to the surface reservoirs and the removal of an equal volume of water by evaporation.

The problem is to calculate the steady-state concentration of dissolved phosphate in the five oceanic reservoirs, assuming that 95 percent of all the phosphate carried into each surface reservoir is consumed by plankton and carried downward in particulate form into the underlying deep reservoir (Figure 3–2). The remaining 5 percent of the incoming phosphate is carried out of the surface reservoir still in solution. Nearly all of the phosphorus carried into the deep sea in particles is restored to dissolved form by consumer organisms. A small fraction—equal to 1 percent of the original flux of dissolved phosphate into the surface reservoir—escapes dissolution and is removed from the ocean into seafloor sediments. This permanent removal of phosphorus is balanced by a flux of dissolved phosphate in river water, with a concentration of 10^{-3} mole P/m^3.

3.3 Simultaneous Algebraic Equations

The first step in solving this problem is to write down a series of equations, one for each reservoir, that expresses the balance between the phosphorus supplied to each reservoir (the source) and the phosphorus removed from each reservoir (the sink). Corresponding to each of the arrows in Figure 3–1 is a flux of phosphorus equal to the water flux multiplied by the concentration of phosphorus in the reservoir from which the water is flowing. Let *sat* be the concentration of phosphate in the surface Atlantic, *sind* in the surface Indo-Pacific, *dind* in the deep Indo-Pacific, *ant* in the Antarctic, and *dat* in the deep Atlantic. Express all concentrations in units of 10^{-3} mole P/m^3.

Consider the balance in the shallow Atlantic reservoir. The flux of phosphorus is

$$8 * dat + 12 * ant + 0.25 * 1$$

Of this flux, 95 percent is removed in particulate form, and 5 percent is carried away in solution (Figure 3–2). The dissolved flux out of the surface Atlantic reservoir is 20 * *sat,* and so the material balance for this reservoir is expressed by the equation

$$(8 * dat + 12 * ant + 0.25 * 1) * (1 - 0.95) = 20 * sat$$

The corresponding equation for the surface Indo-Pacific reservoir is

$$(24 * dind + 0.75 * 1) * 0.05 = 24 * sind$$

and for the deep Indo-Pacific reservoir

$$32 * ant + (24 * dind + 0.75 * 1) * 0.94 = (8 + 24) * dind$$

where the term multiplied by 0.94 represents the dissolution of particulate phosphorus carried into the deep ocean from the surface, and the missing 1 percent represents the flux out of the ocean into sediments, balanced overall by the river-borne flux into the ocean.

The equations for the remaining two reservoirs, Antarctic and deep Atlantic, are

$$8 * dind + 16 * dat + 24 * sind = (12 + 4 + 32) * ant$$
$$20 * sat + 4 * ant + (8 * dat + 12 * ant + 0.25 * 1) * 0.94$$
$$= (8 + 16) * dat$$

The complete system consists of five linear algebraic equations in the five unknown concentrations. The order of the terms in these equations can be rearranged so that the first term in each equation is a number (which may be 0) times *sat*, the second is a number times *sind*, the third is a number times *dind*, the fourth is a number times *ant*, the fifth is a number times *dat*, and the constant terms, which correspond in this system to the river-borne source, appear on the right-hand sides of the equations. After this rearrangement, the equations become

$-20 * sat$	$+0 * sind$	$+0 * dind$	$+0.6 * ant$	$+0.4 * dat$	$= -0.0125$
$0 * sat$	$-24 * sind$	$+1.2 * dind$	$+0 * ant$	$+0 * dat$	$= -0.0375$
$0 * sat$	$+0 * sind$	$-9.44 * dind$	$+32 * ant$	$+0 * dat$	$= -0.705$
$0 * sat$	$+24 * sind$	$+8 * dind$	$-48 * ant$	$+16 * dat$	$= 0$
$20 * sat$	$+0 * sind$	$+0 * dind$	$+15.28 * ant$	$-16.48 * dat$	$= -0.235$

For convenience in notation and computation this system can be represented by an array of coefficients, which I call *sleq*:

-20	0	0	0.6	0.4	-0.0125
0	-24	1.2	0	0	0.0375
0	0	-9.44	32	0	0.705
0	24	8	-48	16	0
20	0	0	15.28	-16.48	-0.235

about which it is understood that each column multiplies a different unknown, that the last column lists the constants on the right-hand sides of the equations, and that the rows represent the five equations that express balance in the five oceanic reservoirs.

The solution begins with a method called *Gaussian elimination* that converts the elements on the diagonal to 1 and the elements below the diagonal to 0. Once this has been done, the solution can be calculated directly by a method called *back substitution*. The required changes in the matrix are possible because the solution will not be altered if an equation is multiplied by a constant or if two equations are added together.

Divide the first row by the first element in that row to get a one on the diagonal. Multiply the first row by the first element in the second row (in this example this element is zero), and subtract the row that results from the second row to get a zero in the first column of the second row. Multiply the first row by the first element in the third row, and subtract the row that results from the third row to get a zero in the first column of the third row. Repeat the process for the fourth and fifth rows. The array now has a one and four zeros in the first column.

1	0	0	−0.03	−0.02	0.000625
0	−24	1.2	0	0	−0.0375
0	0	−9.44	32	0	−0.705
0	24	8	−48	16	0
0	0	0	15.88	−16.08	−0.2475

Now leave the first row alone. Divide the second row by its second element to get a one on the diagonal. Multiply the second row by the second element in the third row, and subtract the row that results from the third row to get a zero in the second element of the third row. Repeat for the fourth and fifth rows.

1	0	0	−0.03	−0.02	0.000625
0	1	−0.05	0	0	0.0015625
0	0	−9.44	32	0	−0.705
0	0	9.2	−48	16	−0.0375
0	0	0	15.88	−16.08	−0.2475

Now leave the second row alone. Divide the third row, and subtract from the fourth and fifth rows.

1	0	0	−0.03	−0.02	0.000625
0	1	−0.05	0	0	0.0015625
0	0	1	−3.389831	0	0.0746822
0	0	0	−16.81356	16	−0.7245763
0	0	0	15.88	−16.08	−0.2475

Divide the fourth row, and subtract from the fifth row.

1	0	0	−0.03	−0.02	0.000625
0	1	−0.05	0	0	0.0015625
0	0	1	−3.389831	0	0.0746822
0	0	0	1	−0.9516129	0.0430947
0	0	0	0	−0.9683867	−0.9318448

Divide the fifth row by its first nonzero element.

1	0	0	−0.03	−0.02	0.000625
0	1	−0.05	0	0	0.0015625
0	0	1	−3.389831	0	0.0746822
0	0	0	1	−0.9516129	0.0430947
0	0	0	0	1	0.9622653

Gaussian elimination has converted the original system of equations into the following system:

$$u(1) + sleq(1,2)*u(2) + sleq(1,3)*u(3) + sleq(1,4)*u(4) + sleq(1,5)*u(5) = sleq(1,6)$$
$$u(2) + sleq(2,3)*u(3) + sleq(2,4)*u(4) + sleq(2,5)*u(5) = sleq(2,6)$$
$$u(3) + sleq(3,4)*u(4) + sleq(3,5)*u(5) = sleq(3,6)$$
$$u(4) + sleq(4,5)*u(5) = sleq(4,6)$$
$$u(5) = sleq(5,6)$$

in which the values of the *sleq* coefficients have changed. The values of the unknowns can now be readily calculated by the method of back substitution. The last equation gives the last unknown directly, which is substituted into the fourth equation to give the value of the fourth unknown. The values of $u(5)$ and $u(4)$ are substituted into the third equation to get the value of $u(3)$; all three are substituted into the second equation to get $u(2)$; and then all four are substituted into the first equation to get $u(1)$. Here are the results:

$sat = u(1)$	$sind = u(2)$	$dind = u(3)$	$ant = u(4)$	$dat = u(5)$
0.04863429	0.167805	3.324849	0.9587993	0.9622657

Program DGC03 carries out the calculation.

You may have noticed that the calculation I have outlined will fail if there is a zero on the diagonal. That is, division by zero is not allowed. And so to be on the safe side, the program checks for zero before dividing by the element on the diagonal. It is easy enough, however, to deal with a zero if it occurs. In subroutine SWAPPER, the program simply switches two equations (exchanging rows in the array) in order to put a nonzero

```
'Program DGC03 solves the steady state ocean model using
'Gaussian elimination and back substitution.
READ nrow                'number of equations
ncol = nrow + 1          'number of columns
DIM sleq(nrow, ncol), unk(nrow)
FOR jrow = 1 TO nrow
    FOR jcol = 1 TO ncol
        READ sleq(jrow, jcol)      'coefficients
    NEXT jcol
NEXT jrow
'================================================================
DATA 5
DATA -20, 0, 0, .6, .4, -.0125
DATA 0, -24, 1.2, 0, 0, -.0375
DATA 0, 0, -9.44, 32, 0, -.705
DATA 0, 24, 8, -48, 16, 0
DATA 20, 0, 0, 15.28, -16.48, -.235
'================================================================
GOSUB GAUSS
FOR jrow = 1 TO nrow
    PRINT unk(jrow);
NEXT jrow
PRINT
END
'****************************************************************************
GAUSS: 'Subroutine GAUSS solves a system of simultaneous linear algebraic
'equations by Gaussian elimination and back substitution.  The number of
'equations (equal to the number of unknowns) is NROW.  The coefficients
'are in array SLEQ(NROW,NROW+1), where the last column is the constants
'on the right hand sides of the equations.
'The answers are returned in the array UNK(NROW).
FOR jrow = 1 TO nrow
    diag = sleq(jrow, jrow)
    IF diag = 0 THEN GOSUB SWAPPER          'check for zero on the diagonal
    FOR jcol = jrow TO ncol
        'Divide by coefficient on the diagonal
        sleq(jrow, jcol) = sleq(jrow, jcol) / diag
    NEXT jcol
    FOR jr = jrow + 1 TO nrow
        coeff1 = sleq(jr, jrow)
        'Zeroes below the diagonal
        FOR jcol = jrow TO ncol
```

```
                 sleq(jr, jcol) = sleq(jr, jcol) - sleq(jrow, jcol) * coeff1
          NEXT jcol
       NEXT jr
NEXT jrow
'Calculate unknowns by back substitution
unk(nrow) = sleq(nrow, ncol)
FOR jrow = nrow - 1 TO 1 STEP -1
    rsum = 0
    FOR jcol = jrow + 1 TO ncol - 1
        rsum = rsum + unk(jcol) * sleq(jrow, jcol)
    NEXT jcol
    unk(jrow) = sleq(jrow, ncol) - rsum
NEXT jrow
RETURN
'****************************************************************************
SWAPPER:      'This subroutine exchanges rows to get a zero coefficient
'off the diagonal
FOR jr = jrow + 1 TO nrow
    IF sleq(jr, jrow) = 0 THEN 2010
    FOR jcol = jrow TO ncol                   'interchange rows
        temp = sleq(jr, jcol)
        sleq(jr, jcol) = sleq(jrow, jcol)
        sleq(jrow, jcol) = temp
    NEXT jcol
    diag = sleq(jrow, jrow)
    RETURN
2010 NEXT jr
PRINT "ZERO ELEMENT ON THE DIAGONAL."
PRINT "CAN NOT SOLVE SIMULTANEOUS EQUATIONS."
STOP
RETURN
```

element on the diagonal. The answers are not affected by a change in the order of the equations. So if there is a zero on the diagonal, the program will search down the column below it to find a nonzero element. Then it will exchange the two rows and proceed as before. If, however, all of the elements in a column are zero, then the system cannot be solved, because there are, in effect, more unknowns than equations. The program will print an error message and stop if this condition occurs.

3.4 Simultaneous Differential Equations

The steady-state problem yields a system of simultaneous linear algebraic equations that can be solved by Gaussian elimination and back substitution. I shall turn now to calculating the time evolution of this system, starting from a phosphate distribution that is not in steady state. In this calculation, assume that the phosphate concentration is initially the same in all reservoirs and equal to the value in river water, 10^{-3} mole P/m^3. How do the concentrations evolve from this starting value to the steady-state values just calculated?

The phosphate content of each reservoir is the volume of that reservoir multiplied by the concentration of phosphate in that reservoir. Volumes are constant, so the rate of change of the content is the volume multiplied by the rate of change of the concentration. But for each reservoir, the rate of change of the content is the sum of all the fluxes of phosphorus in (the sources) minus the sum of all the fluxes of phosphorus out (the sinks). Thus, the content of the first reservoir, Atlantic surface, changes at the rate

$$v1 \ \frac{dy1}{dt} \ = \ ex11 * y1 + ex12 * y2 + ex13 * y3 + ex14 * y4$$
$$4x19 + ex15 * y5 - ex16$$

where $v1$ is the volume of reservoir *1*, $y1$ is the concentration, and the ex values are the rates of flow of water from one reservoir to another (see Section 3.2). The values of the ex coefficients are repeated here. The units are Sverdrups, where 1 Sverdrup = 10^6 m^3/sec = 3.16×10^{13} m^3/y. In this notation, the first number refers to the row, and the second number to the column of this array.

sat	-20	0	0	0.6	0.4	-0.0125
sind	0	-24	1.2	0	0	-0.0375
dind	0	0	-9.44	32	0	-0.705
ant	0	24	8	-48	16	0
dat	20	0	0	15.28	-16.48	-0.235

I divide both sides of the equation by the volume to obtain an equation for the rate of change of concentration:

$$\frac{dy1}{dt} = (ex11 * y1 + ex12 * y2 + ex13 * y3 + ex14 * y4 + ex15 * y5 - ex16)/v1$$

The equations for the other reservoirs have exactly the same form:

$$\frac{dy2}{dt} = (ex21 * y1 + ex22 * y2 + ex23 * y3 + ex24 * y4 + ex25 * y5 - ex26)/v2$$

$$\frac{dy3}{dt} = (ex31 * y1 + ex32 * y2 + ex33 * y3 + ex34 * y4 + ex35 * y5 - ex36)/v3$$

$$\frac{dy4}{dt} = (ex41 * y1 + ex42 * y2 + ex43 * y3 + ex44 * y4 + ex45 * y5 - ex46)/v4$$

$$\frac{dy5}{dt} = (ex51 * y1 + ex52 * y2 + ex53 * y3 + ex54 * y4 + ex55 * y5 - ex56)/v5$$

The volumes, expressed in 10^{17} m^3, are

$$v1 = 0.3, \ v2 = 0.9, \ v3 = 8.1, \ v4 = 1.5, \ v5 = 2.7$$

(Broecker and Peng, 1982, p. 382).

The problem is to calculate new values of the variables, y, at a new time, $x + delx$, given the known values of y at time x. In the reverse Euler method described in Section 2.4,

$$y(x + delx) = y(x) + dely$$

where

$$dely = delx * yp(x + delx, y(x + delx))$$

and yp is the rate of change of y evaluated at $x + delx$. The equations for yp were just given. Each rate of change is, in general, a function of all of the variables, y, and substitution in each of these equations of $y + dely$ for y yields

$dely1/delx = (ex11 * (y1 + dely1) + ex12 * (y2 + dely2) + ex13 * (y3 + dely3) + ex14 * (y4 + dely4) + ex15 * (y5 + dely5) - ex16)/v1$

$dely2/delx = (ex21 * (y1 + dely1) + ex22 * (y2 + dely2) + ex23 * (y3 + dely3) + ex24 * (y4 + dely4) + ex25 * (y5 + dely5) - ex26)/v2$

$dely3/delx = (ex31 * (y1 + dely1) + ex32 * (y2 + dely2) + ex33 * (y3 + dely3) + ex34 * (y4 + dely4) + ex35 * (y5 + dely5) - ex36)/v3$

$dely4/delx = (ex41 * (y1 + dely1) + ex42 * (y2 + dely2) + ex43 * (y3 + dely3) + ex44 * (y4 + dely4) + ex45 * (y5 + dely5) - ex46)/v4$

$dely5/delx = (ex51 * (y1 + dely1) + ex52 * (y2 + dely2) + ex53 * (y3 + dely3) + ex54 * (y4 + dely4) + ex55 * (y5 + dely5) - ex56)/v5$

The equations can now be rearranged to give a system of simultaneous linear algebraic equations for the unknown values *dely* in terms of the known values of $y(x)$ and the *ex* coefficients:

$(ex11/v1 - 1/delx) * dely1 + (ex12 * dely2 + ex13 * dely3 + ex14 * dely4 + ex15 * dely5)/v1 = (ex16 - (ex11 * y1 + ex12 * y2 + ex13 * y3 + ex14 * y4 + ex15 * y5))/v1$

$ex21 * dely1/v2 + (ex22/v2 - 1/delx) * dely2 + (ex23 * dely3 + ex24 *$
    $dely4 + ex25 * dely5)/v2 = (ex26 - (ex21 * y1 + ex22 * y2 + ex23 *$
    $y3 + ex24 * y4 + ex25 * y5))/v2$

$(ex31 * dely1 + ex32 * dely2)/v3 + (ex33/v3 - 1/delx) * dely3 + (ex34 *$
    $dely4 + ex35 * dely5)/v3 = (ex36 - (ex31 * y1 + ex32 * y2 + ex33 *$
    $y3 + ex34 * y4 + ex35 * y5))/v3$

$(ex41 * dely1 + ex42 * dely2 + ex43 * dely3)/v4 + (ex44/v4 - 1/delx) *$
    $dely4 + ex45 * dely5/v4 = (ex46 - (ex41 * y1 + ex42 * y2 + ex43 * y3$
    $+ ex44 * y4 + ex45 * y5))/v4$

$(ex51 * dely1 + ex52 * dely2 + ex53 * dely3 + ex54 * dely4)/v5$
    $+ (ex55/v5 - 1/delx) * dely5 = (ex56 - (ex51 * y1 + ex52 * y2$
    $+ ex53 * y3 + ex54 * y4 + ex55 * y5))/v5$

These manipulations are not as complicated as they may at first appear, for I have written out the expressions in full detail in order to avoid possible uncertainty about just what the manipulations are. Note that the right-hand sides of these equations are the steady-state equations solved in Section

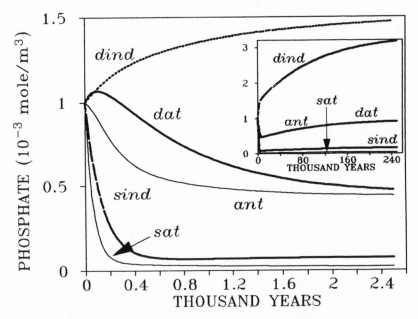

Fig. 3–3. Evolution of phosphate concentrations in the different oceanic reservoirs. The time step for the initial adjustment is 25 years. For the long-term evolution shown in the insert, a time step of 2500 years was used.

```
'Program DGC04 solves the time-dependent ocean simulation by the reverse
'Euler method using Gaussian elimination and back substitution.
nrow = 5                'the number of equations and unknowns
ncol = nrow + 1
DIM sleq(nrow, ncol), unk(nrow), excoeff(nrow, ncol), ovol(nrow)
DIM y(nrow), dely(nrow)
x = 0                   'starting time
delx = 2500             'time step
nstep = 100             'number of steps
GOSUB CORE
END
'*********************************************************************
CORE:   'Subroutine that directs the work
GOSUB SPECS
OPEN "RESULTS.PRN" FOR OUTPUT AS #1 'results to a file for plotting
GOSUB PRINTER
FOR nx = 1 TO nstep
    GOSUB EQUATIONS
    GOSUB GAUSS
    FOR jrow = 1 TO nrow
        dely(jrow) = unk(jrow)
    NEXT jrow
    GOSUB STEPPER
NEXT nx
CLOSE #1
RETURN
'*********************************************************************
PRINTER:    'Subroutine writes a file for subsequent plotting
sat = y(1): sind = y(2): dind = y(3): ant = y(4): dat = y(5)
PRINT x; sat; sind; dind; ant; dat
PRINT #1, x; sat; sind; dind; ant; dat
RETURN
'*********************************************************************
EQUATIONS: 'Subroutine to calculate the coefficient matrix, SLEQ
FOR jrow = 1 TO nrow
    tempsum = 0
    FOR jcol = 1 TO nrow
        sleq(jrow, jcol) = excoeff(jrow, jcol)
     .  tempsum = tempsum + excoeff(jrow, jcol) * y(jcol)
    NEXT jcol
    sleq(jrow, jrow) = sleq(jrow, jrow) - 1 / delx
    sleq(jrow, nrow + 1) = excoeff(jrow, nrow + 1) - tempsum
```

```
NEXT jrow
RETURN
'*******************************************************************************
STEPPER:    'Subroutine steps forward in time
FOR jrow = 1 TO nrow
    y(jrow) = y(jrow) + dely(jrow)
NEXT jrow
x = x + delx
GOSUB PRINTER
RETURN
'*******************************************************************************
SPECS: 'Subroutine to read in the specifications of the problem
FOR jrow = 1 TO nrow
    FOR jcol = 1 TO nrow + 1
        READ excoeff(jrow, jcol)  'fluxes between ocean reservoirs
    NEXT jcol
NEXT jrow
'===============================================
DATA -20, 0, 0, .6, .4, -.0125
DATA 0, -24, 1.2, 0, 0, -.0375
DATA 0, 0, -9.44, 32, 0, -.705
DATA 0, 24, 8, -48, 16, 0
DATA 20, 0, 0, 15.28, -16.48, -.235
'===============================================
'Fluxes expressed in 10^6 m^3/s
FOR jrow = 1 TO nrow
    READ ovol(jrow)          'volumes of reservoirs
NEXT jrow
DATA .3, .9, 8.1, 1.5, 2.7
'  Volumes expressed in 10^17 m^3
FOR jrow = 1 TO nrow
    FOR jcol = 1 TO nrow + 1
        excoeff(jrow, jcol) = excoeff(jrow, jcol) * 3.15576E-04
        'convert 10^6 m^3/s to 10^17 m^3/y
        excoeff(jrow, jcol) = excoeff(jrow, jcol) / ovol(jrow)
        'divide the equation by the volume of the reservoir
    NEXT jcol
NEXT jrow
FOR jrow = 1 TO nrow
    y(jrow) = 1              'initial values
NEXT jrow
```

```
RETURN
'***********************************************************************
```

Plus subroutines GAUSS and SWAPPER from Program DGC03

3.3. When the ys have their steady-state values, the right-hand sides of these equations are zero, so the solution will yield zero for all of the *dely*s. The solution of the time-dependent system is related to the solution of the steady-state system.

Program DGC04 solves the time-dependent problem. Subroutine EQUA-TIONS evaluates the coefficients of the unknown *dely*s in the manner just outlined, and then subroutine GAUSS solves for the values of *dely*. Sub-routine STEPPER steps forward in time by incrementing x and y. Subroutine SPECS sets the values of the parameters of the problem, converting units where necessary, and PRINTER writes the results to a file for plotting.

The results calculated with this program are shown in Figure 3–3. All of the concentrations initially were equal to the concentration in river water. In a matter of a few thousand years, phosphorus is redistributed among the oceanic reservoirs, with its concentrations decreasing in surface reservoirs and increasing in the deep ocean. The small reservoirs adjust more rapidly than do the large ones. On a much longer time scale, several hundred thousand years, phosphorus accumulates in the ocean to bring the sedimen-tary sink into balance with the river source of dissolved phosphate. After about half a million years, the time-dependent solution returns the steady-state concentrations previously calculated.

3.5 Summary

The application of the reverse Euler method of solution to a system of coupled differential equations yields a system of coupled algebraic equa-tions that can be solved by the method of Gaussian elimination and back substitution. In this chapter I demonstrated the solution of simultaneous algebraic equations by means of this method and showed how the solution of algebraic equations can be used to solve the related differential equa-tions. In the process, I presented subroutine GAUSS, the computational engine of all of the programs discussed in the chapters that follow.

4 How to Solve a Nonlinear System

4.1 Introduction

In a linear system the expressions, yp, for rates of change are linear functions of the dependent variables, y. More complicated functions of y do not appear, not even products of the dependent variables like $y1 * y2$. But most theoretical problems in Earth system science involve non-linearities. For example, the rate at which a chemical reaction consumes species 1 may be proportional to the product of the concentrations of species 1 and the species 2 with which it is reacting, $y1 * y2$. In this chapter I shall describe and solve a simple nonlinear system involving the reaction between dissolved oxygen and organic carbon in the deep sea.

I shall show how the nonlinear system can be represented by a linear system, provided that changes in the dependent variables are made in sufficiently small increments. Such increments are kept small by stepping forward in time with small steps. The time step can be adjusted automatically during the calculation so as to keep the increments small enough but no smaller than necessary. Steps that are too large cause errors or even instability, and steps that are too small waste time.

The representation of a nonlinear system by its linear equivalent (for small increments) calls on algebraic manipulations that can be tedious and a prolific source of mistakes in complicated systems. This algebra can be avoided, however, by letting the computer perform the equivalent manipulations numerically. I shall demonstrate how to do this, finishing the chapter with a program that can solve coupled nonlinear systems directly from the equations for the rates of change of the dependent variables, automatically adjusting the time step to small values when the rates of change are large and to large values when the rates of change are small.

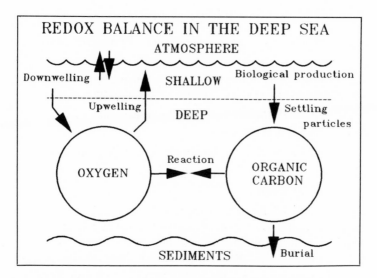

Fig. 4–1. The oxidation–reduction balance of the deep sea.

4.2 The Oxidation–Reduction Balance of the Deep Sea

Figure 4–1 shows a simple model of the processes that control the oxygen balance of the deep sea. Dissolved oxygen in surface seawater equilibrates with the atmosphere. Downwelling currents carry dissolved oxygen into the deep sea, where most of it reacts (metabolically) with organic matter carried into the deep sea in the form of settling particles of biological origin. A small fraction of this carbon escapes oxidation in the deep sea, to be buried in sediments. Oxygen that is not consumed by reaction with organic carbon is carried back to the surface by upwelling currents.

According to this model, the rates of change in the amounts of oxygen and organic carbon in the deep sea can be written as ordinary differential equations:

$$vold * \frac{doxy}{dt} = (so - oxy) * wflux - oxy * corg * koc * 1.3$$

$$\frac{dcorg}{dt} = prod - corg/btime - oxy * corg * koc$$

where, quoting values from Broecker and Peng (1982),

oxy mole/m^3 is the concentration of dissolved oxygen in deep sea-water.

$so = 0.34$ mole/m^3 is the concentration of dissolved oxygen in surface water, determined by equilibrium with the atmosphere.

$wflux = 1000 \times 10^{12}$ m^3/y is the rate of exchange of water between deep and shallow reservoirs.

$vold = 1.23 \times 10^{18}$ m^3 is the volume of the deep-sea reservoir.

$corg$ is the amount of reactive organic carbon in the deep sea expressed in units of 10^{18} moles.

koc is the rate coefficient for the reaction between dissolved oxygen and reactive organic carbon. I assume that the rate of this reaction is proportional to the product of the concentration of dissolved oxygen and the amount of reactive organic carbon. The units are m^3 mole^{-1} y^{-1}. The factor of 1.3 is the ratio of oxygen to carbon in the respiration reaction, which differs from 1 because the organic nitrogen is oxidized at the same time as is the organic carbon.

$prod$ is the rate of transfer of organic matter into the deep sea expressed in units of 10^{12} mole/y.

$btime$ (years) is the time that organic matter can stay in the deep sea and on the bottom before it is buried in sediments and removed from the region of chemical interaction.

Reasonable trial values of the parameters are

$$koc = 1 \text{ m}^3 \text{ mole}^{-1} \text{ y}^{-1}$$
$$prod = 100 \times 10^{12} \text{ mole/y}$$
$$btime = 1000 \text{ y}$$

For starting conditions I shall use an ocean initially lifeless; in other words, at $x = 0$, $oxy = so$ and $corg = 0$.

4.3 Small Increments Make the System Linear

My first attempt to calculate the time history of a geochemical system (Section 2.3) used the obvious approach (the direct Euler method) of evaluating the time derivatives and stepping forward. But it was not successful because the solution was unstable for larger time steps. The instability arose when the calculation diverged from the correct result, tried to get back, and went too far. The reverse Euler method (Section 2.4) avoids this problem by evaluating the time derivative at the future time, not the present time. Accordingly, instead of the unstable

$$y(x + delx) = y(x) + yp(x) * delx$$

use

$$y(x + delx) = y(x) + yp(x + delx) * delx$$

where y is a dependent variable, x the independent variable, yp the slope, and *delx* the increment in x (time step). This approach turns out to be nearly always stable. The difficulty, of course, is that $y(x + delx)$ is not known in advance, and so $yp(x + delx)$ is not known either. A system of algebraic equations thus must be solved.

Define the increment in y,

$$dely = y(x + delx) - y(x) = yp(x + delx) * delx$$

In the expression for yp, substitute $y(x + delx) = y(x) + dely$ to derive a system of algebraic equations for *dely*. Here are the equations presented earlier:

$$yp(1) = (so - y(1)) * wflux/vold - y(1) * y(2) * koc * 1.3/vold$$
$$yp(2) = prod - y(2)/btime - y(1) * y(2) * koc$$

To clarify the notation I will write the indexes outside the parentheses; thus

$$
\begin{aligned}
dely1/delx &= yp1(x + delx) \\
&= (so - y1(x) - dely1) * wflux/vold - (y1(x) * y2(x) * koc * 1.3 \\
&\quad - y1(x) * dely2 * koc * 1.3 - dely1 * y2(x) * koc * 1.3 \\
&\quad - dely1 * dely2 * koc * 1.3)/vold
\end{aligned}
$$

$$
\begin{aligned}
dely2/delx &= yp2(x + delx) \\
&= prod - (y2(x) + dely2)/btime \\
&\quad - y1(x) * y2(x) * koc - y1(x) * dely2 * koc \\
&\quad - dely1 * y2(x) * koc - dely1 * dely2 * koc
\end{aligned}
$$

$y1(x)$ and $y2(x)$ are the current values of the dependent variables; they are known. Although this is a system of algebraic equations for the unknowns, *dely1* and *dely2*, it is a nonlinear system because the last term in each equation contains the product of the unknowns. I make the equations linear by dropping the last term in each equation on the grounds that *dely* is much smaller than y, and so the product of *dely1* and *dely2* is negligibly small.

Then this system of simultaneous linear algebraic equations can be solved using the subroutine GAUSS developed in Section 3.3. Because I have dropped the nonlinear term, I must always use a *delx* sufficiently small to ensure that all the *dely* values are indeed much smaller than the *y* values.

The system of equations therefore becomes

$$dely1 * (1/delx + wflux/vold + y2 * koc * 1.3/vold) + dely2 * y1 * koc * 1.3/vold$$
$$= (so - y1) * wflux/vold - y1 * y2 * koc * 1.3/vold$$
$$dely1 * (y2 * koc) + dely2 * (1/delx + 1/btime + y1 * koc)$$
$$= prod - y2/btime - y1 * y2 * koc$$

Program DGC05 implements this solution with a subroutine to evaluate the coefficients of the *sleq* matrix. It calls subroutine GAUSS to calculate the increments *dely,* and then it steps forward in time and repeats.

The results of the calculation are plotted in Figure 4–2. Dissolved oxygen levels out at a value of about 0.2 mole/m³. The average length of time

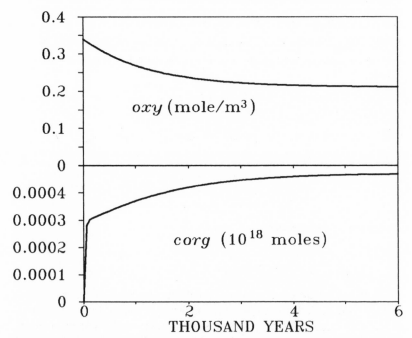

Fig. 4–2. The evolution of dissolved oxygen and reactive organic carbon in the deep sea. The calculations used a time step *delx* = 60 years. A shorter step is needed at the beginning, when the change is particularly rapid.

```
'Program DGC05 solves the time-dependent simulation of the deep sea redox
'balance in linearized form by the reverse Euler method
'using Gaussian elimination and back substitution.
nrow = 2                'the number of equations and unknowns
ncol = nrow + 1
DIM sleq(nrow, ncol), unk(nrow)
DIM y(nrow), dely(nrow)
x = 0                   'starting time
delx = 60               'time step
nstep = 100             'number of steps
GOSUB CORE
END
'****************************************************************************
PRINTER:   'Subroutine writes a file for subsequent plotting
oxy = y(1): corg = y(2)
PRINT x; oxy; corg
PRINT #1, x; oxy; corg
RETURN
'****************************************************************************
EQUATIONS: 'Subroutine to calculate the coefficient matrix, SLEQ
sleq(1, 1) = 1 / delx + wflux / vold + y(2) * koc * 1.3 / vold
sleq(1, 2) = y(1) * koc * 1.3 / vold
sleq(1, 3) = (so - y(1)) * wflux / vold - y(1) * y(2) * koc * 1.3 / vold
sleq(2, 1) = y(2) * koc
sleq(2, 2) = 1 / delx + 1 / btime + y(1) * koc
sleq(2, 3) = prod - y(2) / btime - y(1) * y(2) * koc
RETURN
'****************************************************************************
SPECS: 'Subroutine to read in the specifications of the problem
so = .34         'mole/m^3
wflux = .001     '10^18 m^3/y
vold = 1.23      '10^18 m^3
koc = 1          'm^3/mole/y
btime = 1000     'y
prod = .0001     '10^18 mole/y
oxy = so
corg = 0
y(1) = oxy                  'initial values
y(2) = corg
RETURN
```

'**
Plus subroutines GAUSS and SWAPPER from Program DGC03
Subroutines CORE and STEPPER from Program DGC04

that reactive organic carbon survives in the deep sea before being oxidized is $1/(koc * oxy) = 5$ years. The carbon amount levels out at about 0.0005×10^{18} mole. The rate of burial of carbon in sediments is $corg/btime = 0.5 \times 10^{12}$ mole/y, which is 0.005 times the rate at which carbon settles into the deep sea from the shallow sea, $prod = 100 \times 10^{12}$ mole/y. These results are reasonable for the modern ocean (Broecker and Peng, 1982), and of course, I chose the values of *koc, btime,* and *prod* to yield reasonable results. *Tuning* is the name given to this process of choosing values for the parameters of a simulation so that it yields reasonable results. Tuning is justified because these parameters are frequently not well known; the simulation is in any event a fairly simple representation of a complex system; and a simulation must yield reasonable results if it is to be useful.

The rate of change at the beginning of the run, particularly of *corg,* is so great that a time step of 60 years is too large. The first few steps yield corners in the plot of *corg* against time. Figure 4–3 shows how a time step

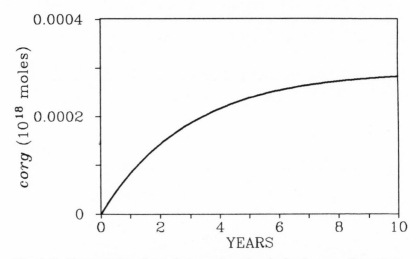

Fig. 4–3. The evolution of reactive organic carbon in the deep sea. The calculations used a time step of 0.1 years. The rapid initial change of *corg* is now resolved, and the change in *oxy* during a period of only 10 years is negligible.

of 0.1 y smooths out the corners, adequately resolving the rapid initial evolution of *corg*. The time step must be small enough to permit approximation by finite difference *dely/delx* of the derivative dy/dx and also to permit the linearization of a nonlinear system by the neglect of terms involving the product of *dely*s. Later in the run, on the other hand, the rates of change are small, and so it would be efficient to use a larger time step. A method of adjusting the time step during a calculation would be convenient.

4.4 How to Adjust the Time Step

When deciding whether to adjust the time step, I compare the relative increment in each of the dependent variables ABS(*dely/y*) with a predetermined maximum permissible value, which I call *maxinc*. Subroutine CHECKSTEP in Program DGC06 performs the manipulations described in this section. There are two reasons that *dely* must be small. One is the finite difference approximation *dely/delx* for the derivative dy/dx. The other is that I have linearized the algebraic equations by neglecting products of *dely*. A reasonable value for *maxinc* is probably 0.1, which would make the nonlinear terms less than 1 percent of terms like $y1 * y2$. For a more accurate solution I would make *maxinc* smaller. To speed up the calculation, I would make *maxinc* larger. When calculating the relative increment *dely/y*, I test for the possibility that y is zero, for division by zero is to be avoided. If y is zero, I will set the relative increment for that equation to *maxinc*/1.4. It will have no effect on what happens next, and y will presumably not be zero for long. I set *mstep* as an upper limit on *delx* to guard against overflow (a value of *delx* too large for the computer to remember or to manipulate).

If any of the relative increments are larger than *maxinc*, I will divide *delx* by 2 and repeat the time step. I will not plot the results or go on to the next time step until *delx* is small enough. If all of the relative increments are less than *maxinc*, then *delx* is small enough; the program will record the results and go on to the next time step. However, if all of the relative increments are smaller than *maxinc*/2, I will increase *delx* before taking the next time step. I will multiply *delx* by 1.5, which leaves a margin of safety. Multiplication by 2 might make it necessary to repeat the next time step and reduce *delx* again. Although more elaborate adjustment schemes are possible, this one is simple and effective.

The results of a calculation with program DGC06 are shown in Figure

```
'Program DGC06 solves the time-dependent simulation of the deep sea redox
'balance in linearized form by the reverse Euler method
'using Gaussian elimination and back substitution.
'With adjustable time step.
nrow = 2                    'the number of equations and unknowns
ncol = nrow + 1
DIM sleq(nrow, ncol), unk(nrow)
DIM y(nrow), dely(nrow)
x = 0
delx = 1
nstep = 300
mstep = 10000               'Maximum time step
maxinc = .1                 'Maximum relative change in dependent variable
GOSUB CORE
END
'****************************************************************************
CORE:    'Subroutine directs the calculation
GOSUB SPECS
OPEN "RESULTS.PRN" FOR OUTPUT AS #1 'results to a file for plotting
GOSUB PRINTER
FOR nx = 1 TO nstep
    GOSUB EQUATIONS
    GOSUB GAUSS
    FOR jrow = 1 TO nrow
        dely(jrow) = unk(jrow)
    NEXT jrow
    GOSUB CHECKSTEP
NEXT nx
CLOSE #1
RETURN
REM *************************************************************************
CHECKSTEP:   'This subroutine identifies the largest relative increment
'ABS(dely/y); compares it with the specified value MAXINC, and adjusts the
'step size if adjustment is needed
biginc = 0
FOR jrow = 1 TO nrow
'Find the largest relative increment or use MAXINC/1.4 if Y is zero
    IF y(jrow) = 0 THEN
        relinc = maxinc / 1.4
    ELSE
        relinc = ABS(dely(jrow) / y(jrow))
```

```
    END IF
    IF biginc < relinc THEN biginc = relinc
    IF biginc = relinc THEN bigone = jrow
NEXT jrow
IF biginc < maxinc THEN 1210
delx = delx / 2                          'reduce DELX and return without
RETURN                                   'stepping forward
1210 GOSUB STEPPER                       'increment Y and X and write file
IF biginc > maxinc / 2 THEN RETURN       'no change needed
delx = delx * 1.5                        'increase DELX
IF delx > mstep THEN delx = mstep        'upper limit on DELX
RETURN
'*********************************************************************
PRINTER:    'Subroutine writes a file for subsequent plotting
oxy = y(1): corg = y(2)
PRINT x; oxy; corg; delx
PRINT #1, x; oxy; corg; delx
RETURN
'*********************************************************************
Plus subroutines GAUSS and SWAPPER from Program DGC03
Subroutine STEPPER from Program DGC04
Subroutines EQUATIONS and SPECS from Program DGC05
```

4–4. This calculation was started with $oxy = so$ as before, but with a large amount of reactive organic carbon in the deep sea, $corg = 1 \times 10^{18}$ moles. The scales on the graph are logarithmic, in order to show the full extent of the changes that occur. To avoid the logarithm of 0, the calculation started at time $x = 1$, and the time step was adjusted automatically by the program in subroutine CHECKSTEP. The values of *delx* are plotted.

The first change in the system is the exhaustion of dissolved oxygen by reaction with the large amount of organic carbon initially in the deep sea. Oxygen is removed in less than 10 years, while the amount of organic carbon is little affected. On a time scale of about a thousand years, dissolved oxygen is replenished by transport from the surface, but the rate of supply of organic carbon by biological production is not fast enough to maintain the large initial value of *corg*. After a few thousand years, therefore, oxygen climbs to about 0.2 mole/m^3, in the process driving down the amount of organic carbon to about 0.0005×10^{18} moles. Oxygen and

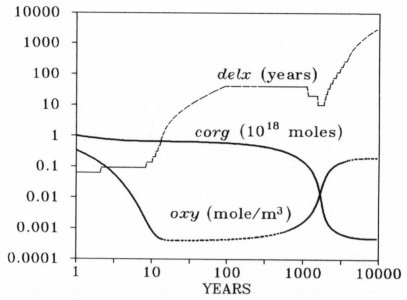

Fig. 4–4. The evolution of dissolved oxygen and reactive organic carbon in the deep sea, starting with a carbon amount much larger than the oxygen amount. The time step *delx* is adjusted automatically, and its values are plotted. The scales are logarithmic.

carbon react with each other, and their product is nearly constant, except when the rates of change are large.

These results illustrate several features of the system that we should note. First, the final steady-state values do not depend on the starting conditions. The final values in both Figure 4–2 and 4–4 are the same, a property that makes it possible to solve these systems by a method as simple as the reverse Euler method. Just as the influence of initial conditions disappears after a while, so also does the influence of any computational inaccuracy. Errors do not grow; instead, they decay away. That is, the system is forgiving, and so current conditions do not depend on remote history. Second, the system exhibits very different response times under different circumstances. In Figure 4–2 it is *corg* that changed markedly in less than 10 years. In Figure 4–4 it is *oxy*. The change after a thousand years is much slower, and both constituents evolve at the same rate. The time step, *delx*, responds well to the changing rates of change of the dependent variables. It starts at less than 0.1 y but climbs to almost 100 y in the middle of the calculation, when *oxy* is steadily increasing and *corg* is decreasing. During the more rapid adjustment when *oxy* is rising above *corg*, *delx* decreases and then recovers to more than 1000 y.

4.5 Let the Machine Do the Work

Although this program is behaving well, the system it describes is still very simple. Even for this system, there was some algebraic manipulation in linearizing the equations, and so it is easy to make mistakes, to get a sign wrong or lose a term, errors that can be hard to detect and locate. Before turning to more complicated systems, therefore, I would like to develop a straightforward method of linearizing the equations and evaluating the terms in the *sleq* array. My goal is a program in which it is necessary only to list the equations for the rates of change of the various quantities, leaving the manipulation of these equations to the computer. Here is how I developed this program.

The system of differential equations is of the form

$$\frac{dy(i)}{dx} = yp(i)$$

where the $yp(i)$ are functions of all of the dependent variables $y(i)$. The elements in the last column of the *sleq* array are just the values of $yp(i)$ at the present x, using the known values of $y(i)$. The other elements describe how rapidly yp changes in response to a change in one of the dependent variables. Specifically, the elements in column j describe the response of yp to changes in $y(j)$, and the diagonal elements contain an additional term of $1/delx$. With this description in mind, I used the following procedure to calculate the elements of the *sleq* array.

I calculated all of the yps using the current values of y. These are the elements in the last column of *sleq*. Then I incremented the first dependent variable $y(1)$ by a small amount, *yinc*, and recalculated all of the yps. I subtracted the new values from the original values and divided the difference by *yinc*. These are the elements for the first column of the *sleq* array. I restored $y(1)$ to its original value and repeated the procedure with $y(2)$ to get the elements for the second column. I did this until all the columns had been evaluated. Finally, I added $1/delx$ to the diagonal elements. These procedures yielded values of the elements of the *sleq* array that are identical to those calculated with the linearized algebraic expressions.

I found it convenient to use $y(i)$ times a parameter *dlny* for *yinc*, with a

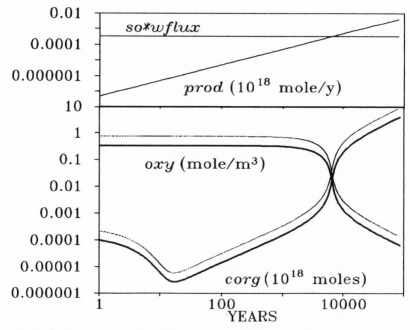

Fig. 4–5. The response of deep-sea oxygen and organic carbon to a linearly increasing productivity. The scales are logarithmic. The solid lines were calculated with program DGC07 using algebraic linearization. The dotted lines, which were moved upward a short way for clarity, were calculated with program DGC08 using automatic linearization. The two programs yield identical results. The top panel compares productivity with the rate of transfer of dissolved oxygen into the deep sea.

check to make sure that *yinc* was not zero (before division by it). These manipulations are implemented by subroutine SLOPER in program DGC08. Figure 4–5 compares the results of calculations using programs DGC07 and DGC08, that is, algebraic linearization versus automatic linearization. The results are identical. Program DGC07 differs from program DGC06 only in the starting values and in the expression for *prod*. I started this calculation at $x = 1$ year with *oxy* = *so* and *corg* = 0.0001×10^{18} moles. The biological productivity was made to increase linearly with time according to

$$prod = prodz * x/2000$$

```
'Program DGC07 is DGC06 with a linearly increasing PROD
nrow = 2                'the number of equations and unknowns
ncol = nrow + 1
DIM sleq(nrow, ncol), unk(nrow), y(nrow), dely(nrow)
x = 1
delx = 1
nstep = 300
mstep = 10000
maxinc = .1
GOSUB CORE
END
'****************************************************************************
EQUATIONS: 'Subroutine to calculate the coefficient matrix
prod = prodz * (x + delx) / 2000
sleq(1, 1) = 1 / delx + wflux / vold + y(2) * koc * 1.3 / vold
sleq(1, 2) = y(1) * koc * 1.3 / vold
sleq(1, 3) = (so - y(1)) * wflux / vold - y(1) * y(2) * koc * 1.3 / vold
sleq(2, 1) = y(2) * koc
sleq(2, 2) = 1 / delx + 1 / btime + y(1) * koc
sleq(2, 3) = prod - y(2) / btime - y(1) * y(2) * koc
RETURN
'****************************************************************************
SPECS: 'Subroutine to read in the specifications of the problem
so = .34        'mole/m^3
wflux = .001    '10^18 m~3/y
vold = 1.23     '10^18 m^3
koc = 1         'm^3/mole/y
btime = 1000    'y
prodz = .0001   '10^18 mole/y
oxy = so
corg = 0
y(1) = oxy                  'initial values
y(2) = corg
RETURN
'****************************************************************************
Plus subroutines GAUSS and SWAPPER from Program DGC03
Subroutine STEPPER from Program DGC04
Subroutines CHECKSTEP, CORE, and PRINTER from Program DGC06
```

```
'Program DGC08 solves redox balance of the deep sea with a linearly
'increasing PROD and automatic linearization of the equations
nrow = 2                 'the number of equations and unknowns
ncol = nrow + 1
DIM sleq(nrow, ncol), unk(nrow), y(nrow), dely(nrow), yp(nrow)
x = 1
delx = 1
nstep = 300
mstep = 10000
maxinc = .1
dlny = .001        'relative increment in Y for linearization in SLOPER
GOSUB CORE
END
'****************************************************************************
SLOPER: REM Subroutine to calculate the coefficient matrix, SLEQ
GOSUB EQUATIONS                          'calculate the derivatives
ncol = nrow + 1
FOR jrow = 1 TO nrow
    sleq(jrow, ncol) = yp(jrow)
NEXT jrow
FOR jcol = 1 TO nrow
    yinc = y(jcol) * dlny
    IF yinc = 0 THEN yinc = dlny       'in case y(jcol)=0
    y(jcol) = y(jcol) + yinc
    GOSUB EQUATIONS
    FOR jrow = 1 TO nrow         'differentiate with respect to Y(JCOL)
        sleq(jrow, jcol) = -(yp(jrow) - sleq(jrow, ncol)) / yinc
    NEXT jrow
    y(jcol) = y(jcol) - yinc     'restore the original value
    sleq(jcol, jcol) = sleq(jcol, jcol) + 1 / delx
        'extra term in diagonal elements
NEXT jcol
RETURN
'****************************************************************************
CORE:    'Subroutine directs the calculation
GOSUB SPECS
OPEN "RESULTS.PRN" FOR OUTPUT AS #1 'results to a file for plotting
GOSUB PRINTER
FOR nx = 1 TO nstep
    GOSUB SLOPER
    GOSUB GAUSS
```

```
    FOR jrow = 1 TO nrow
        dely(jrow) = unk(jrow)
    NEXT jrow
    GOSUB CHECKSTEP
NEXT nx
CLOSE #1
RETURN
'****************************************************************************
Plus subroutines GAUSS and SWAPPER from Program DGC03
Subroutine STEPPER from Program DGC04
Subroutines CHECKSTEP and PRINTER from Program DGC06
Subroutines EQUATIONS and SPECS from Program DGC07
```

where

$$prodz = 0.0001 \times 10^{18} \text{ mole/y}$$

Note that *prod* must be evaluated at the future time $x + delx$ and not at x, because this is the reverse Euler method of calculation, not the direct one.

The results show an initial rapid adjustment of *corg* to the small starting productivity followed by a gradual increase in *corg* until *prod* becomes larger than the rate of transfer of dissolved oxygen to the deep sea. At this point, *oxy* decreases rapidly while *corg* increases.

4.6 Summary

I presented a group of subroutines—CORE, CHECKSTEP, STEPPER, SLOPER, GAUSS, and SWAPPER—that can be used to solve diverse theoretical problems in Earth system science. Together these subroutines can solve systems of coupled ordinary differential equations, systems that arise in the mathematical description of the history of environmental properties. The systems to be solved are described by subroutines EQUATIONS and SPECS. The systems need not be linear, as linearization is handled automatically by subroutine SLOPER. Subroutine CHECKSTEP ensures that the time steps are small enough to permit the linear approximation. Subroutine PRINTER simply preserves during the calculation whatever values will be needed for subsequent study.

The applications that I have considered so far have been limited to simple descriptions of aspects of ocean chemistry, but the powerful computational tools now available can be applied to more complicated problems involving changing ocean or atmospheric composition, isotopic evolution, climate, heat flow, diffusion, or diagenesis. Steady-state problems can be solved as easily as can time-dependent problems, simply by allowing time-dependent calculations with constant parameters to evolve to the steady state. The tools are powerful, but not too hard to understand or to use, as I have sought to demonstrate here.

5 The Carbon System and Several Useful Procedures

5.1 Introduction

The routines developed in previous chapters can be used to simulate a variety of interesting systems in geochemical dynamics and global change. In addition to these routines are devices and procedures that can make easier the process of developing and debugging a simulation. I shall present several such procedures in this chapter, including the management of input and output files, particularly files of starting values, the definition of mnemonic names for variables, a graphic subroutine that provides a runtime view of the progress of a calculation, and the specification of complicated histories by means of a table. These are procedures that I find helpful, but because working with a small computer is a personal matter, you may not find them helpful. By all means, develop your own procedures or modify mine.

As an application of these computational helpers I shall also introduce the carbon system and the equilibrium relationships among the species of carbon dissolved in natural waters.

5.2 Carbonate Equilibria

Carbon dioxide plays a key role in climate, in biological processes, in weathering reactions, and in marine chemistry. I shall next describe how the partial pressure of this gas in the atmosphere may be calculated. Because there is a rapid exchange of carbon dioxide between ocean and atmosphere, we must consider the fate of dissolved carbon.

Carbon dissolved in seawater takes part in fast chemical reactions involving the species dissolved carbon dioxide H_2CO_3, bicarbonate ions

HCO_3^-, and carbonate ions $CO_3^=$. The concentrations of these species are governed by equilibrium relationships (Broecker and Peng, 1982).

$$H_2CO_3 = HCO_3^- + H^+$$
$$k1 = h * hco3/h2co3$$

where *k1* is an equilibrium constant and *h, hco3,* and *h2co3* denote concentrations.

$$HCO_3^- = CO_3^= + H^+$$
$$k2 = h * co3/hco3$$

and after eliminating the hydrogen ion concentration,

$$h2co3 = k2/k1 * hco3^2/co3$$

The partial pressure of gaseous carbon dioxide in equilibrium with dissolved carbon dioxide is given by

$$pco2 = a * h2co3 = kco2 * hco3^2/co3$$

where *kco2 = a * k2/k1*.

The concentrations of the different carbon species depend on total dissolved carbon and the requirement that the solution be electrically neutral, that is, that there be as many negative charges per unit volume as there are positive charges. The total dissolved carbon concentration (frequently called *sigma C*) is

$$sigc = hco3 + co3 + h2co3$$

The total concentration of charge carried by the carbon species is called the *alkalinity* and is given by

$$alk = hco3 + 2 * co3$$

because the bicarbonate ion carries one unit of charge and the carbonate ion carries two. Because the solution must be electrically neutral overall, the alkalinity equals the negative of the total concentration of charge carried by all ions other than bicarbonate and carbonate. Alkalinity is therefore determined by the species other than carbon dissolved in the water. For a given concentration of total dissolved carbon, the proportions as carbonate and bicarbonate ions adjust to yield the required alkalinity.

In the modern ocean, *h2co3* is much smaller than the concentrations of other carbon species, and it can therefore be neglected in the expression for *sigc*. As a good approximation, then,

$$co3 = alk - sigc$$

and

$$hco3 = 2 * sigc - alk$$

There may be situations in the geological past for which this approximation is not valid, however, so I shall use the more complete solution. Substitute

$$h2co3 = kcarb * hco3^2/co3$$

where $kcarb = k2/k1$, then eliminate $co3$, and solve the resulting quadratic equation for $hco3$.

$$hco3 = (sigc - \mathrm{SQRT}(sigc^2 - alk * (2 * sigc - alk) * (1 - 4 * kcarb)))/(1 - 4 * kcarb)$$

and then

$$co3 = (alk - hco3)/2$$

The calculation of the concentrations of dissolved carbon species from total dissolved carbon and alkalinity is carried out in subroutine CARBON-ATE, presented in program DGC09. I have specified the equilibrium constants as functions of water temperature by fitting straight lines to the values tabulated by Broecker and Peng (1982, p. 151).

5.3 Exchange Between Reservoirs

In order to calculate the partial pressure of carbon dioxide, it is necessary to figure the total dissolved carbon and alkalinity as well. I consider three reservoirs—atmosphere, surface sea, and deep sea—as illustrated in Figure 5–1. I distinguish between the concentrations in the surface and deep reservoirs by using a terminal letter s for the surface reservoir and d for the deep reservoir.

The exchange of carbon between atmosphere and surface ocean is characterized by a transfer time *distime*

$$\frac{dpco2}{dt} = (pco2s - pco2)/distime$$

where

$$pco2s = kco2 * hco3s^2/co3s$$

is the partial pressure of carbon dioxide in equilibrium with the carbonate species dissolved in surface seawater. In addition to exchange with the

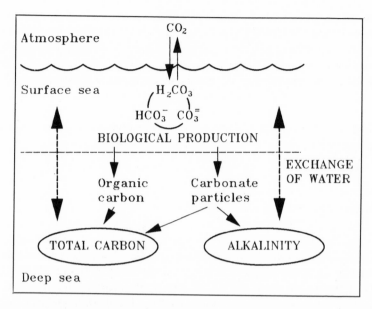

Fig. 5–1. Settling particles of biological origin carry carbon and alkalinity into the deep sea. Carbonate equilibrium reactions in the surface sea affect the atmospheric pressure of carbon dioxide.

atmosphere, the surface ocean exchanges carbon with the deep sea in upwelling and downwelling water. It also loses carbon to the deep sea in the form of settling particles of biological origin. These are partly organic matter, settling at the rate *prod,* and partly calcium carbonate hard parts. The ratio of carbonate to organic carbon is described by the constant *corat.*

$$\frac{\mathrm{d}sigcs}{\mathrm{d}t} = [-(pco2s - pco2)/distime * matmco2 - (1 + corat) * prod$$
$$+ (sigcd - sigcs) * wflux]/vols$$

matmco2 is the mass of carbon dioxide in the atmosphere before the beginning of the human perturbations associated with industrial activity. The carbon dioxide pressure *pco2* is expressed in units of the preindustrial value, and *matmco2* converts these relative units into mass units. Here *pco2* = 1 PAL (preindustrial atmospheric level) corresponds to a partial pressure of 0.00028 bar. The quantity, *vols,* is the volume of the surface sea reservoir, which converts the concentration of a dissolved species into the total mass of that species in the reservoir. In this equation, the term on the left is the rate of change of the concentration of dissolved carbon in the surface ocean; the first term on the right is the exchange of carbon with the atmosphere; the second term is the loss of carbon to the deep ocean as settling particles of organic matter and calcium carbonate; and the third term is the exchange of dissolved carbon with the deep ocean. Finally, *wflux* is the flux of water exchanged between the surface and deep ocean reservoirs.

The net charge of dissolved species other than carbon that fixes the alkalinity is transferred with the exchange of water between surface and deep ocean reservoirs at the rate *wflux*. Charge is also carried from surface to deep in settling particles that dissolve in the deep ocean. Each mole of calcium carbonate carries two equivalent moles of alkalinity because there are two charges on Ca^{2+}. However, each mole of organic carbon is associated with an average of 0.15 moles of organic nitrogen destined to become negatively charged nitrate ions NO_3^-, which subtract from alkalinity. Therefore

$$\frac{dalks}{dt} = [(alkd - alks) * wflux - (2 * corat - 0.15) * prod]/vols$$

The first term on the right is the exchange of water, and the second term is the effect of the settling particles. The contribution of particulate phosphate is negligibly small.

For this calculation I shall assume that all settling organic matter is promptly oxidized in the deep sea to add to the reservoir of total dissolved carbon and that all settling carbonate particles dissolve promptly in the deep sea to contribute both dissolved carbon and alkalinity. I will not allow either organic carbon or carbonate to be preserved in sediments. In fact, a small fraction of the organic matter and a significant fraction of the carbonate particles are incorporated into seafloor sediments, effectively removing carbon and alkalinity from the system. A more realistic simulation of the carbon system would include the accumulation of sediments, with a compensating source of carbon and alkalinity provided by rivers, seafloor weathering, and volcanic processes.

The total dissolved carbon and alkalinity in the deep sea are given by

$$\frac{dsigcd}{dt} = [(1 + corat) * prod - (sigcd - sigcs) * wflux]/vold$$

$$\frac{dalkd}{dt} = [(2 * corat - 0.15) * prod - (alkd - alks) * wflux]/vold$$

In this system, the sum of carbon in all reservoirs and the sum of alkalinity in all reservoirs both remain constant. I am not yet considering possible sources or sinks of carbon and alkalinity.

The starting values and parameters are the following (Broecker and Peng, 1982, p. 69):

$pco2$ = 1 PAL	Atmospheric carbon dioxide
$sigcs$ = 2.01 mole/m^3	Dissolved carbon in surface ocean
$sigcd$ = 2.23 mole/m^3	Dissolved carbon in deep ocean
$alks$ = 2.2 mole/m^3	Alkalinity in surface ocean
$alkd$ = 2.26 mole/m^3	Alkalinity in deep ocean
$distime$ = 8.64 y	Carbon dioxide dissolution time
$matmco2$ = 0.0495 × 10^{18} mole	Atmospheric carbon dioxide mass at 1 PAL
$vols$ = 0.12 × 10^{18} m^3	Volume of surface ocean
$vold$ = 1.23 × 10^{18} m^3	Volume of deep ocean
$prod$ = 0.000175 × 10^{18} mole/y	Rain of particulate organic carbon
$corat$ = 0.25	Ratio of carbonate to organic carbon

Program DGC09 solves this version of the carbon system and is discussed in the next section.

5.4 A Program with Many Refinements

5.4.1 File Management and Starting Values

When I run a calculation with the preceding starting values and the ones in subroutine SPECS of program DGC09, I find that the system adjusts a little because these values do not yield quite a steady state. Because of the nonlinearity, particularly of the expression for $pco2s$, small changes in the total carbon and alkalinity yield noticeable changes in the carbon dioxide partial pressure. For many purposes it is convenient to start with a system in steady state. An easy way to arrive at steady-state starting values is to let a calculation with constant parameters continue long enough to reach a steady

```
'Program DGC09 solves the carbon system in atmosphere, shallow,
'and deep sea.
nrow = 5                'the number of equations and unknowns
ncol = nrow + 1
'==================================================================
DIM sleq(nrow, ncol), unk(nrow), y(nrow), dely(nrow), yp(nrow)
'Establish arrays for graphics routines, GRAFINIT and PLOTTER
numplot = 5             'Number of variables to plot.  Not more than 19
DIM plotz(numplot), plots(numplot), plotl$(numplot), ploty(numplot)
'==================================================================
GOSUB SPECS
GOSUB FILER
xstart = 1              'time to start
xend = 10000           'time to stop, in years
x = xstart
delx = .1
mstep = 10000
GOSUB CORE
END
'*************************************************************************
CORE:   'Subroutine directs the calculation
maxinc = .1
dlny = .001
count = 0               'count time steps
GOSUB GRAFINIT
GOSUB PRINTER
DO WHILE x < xend
    count = count + 1
    GOSUB SLOPER
    GOSUB GAUSS
    FOR jrow = 1 TO nrow
        dely(jrow) = unk(jrow)
    NEXT jrow
    GOSUB CHECKSTEP
LOOP
GOSUB STOPPER
RETURN
'*************************************************************************
DEFINITIONS:    'Subroutine defines dependent variables
pco2 = y(1): sigcs = y(2): sigcd = y(3): alks = y(4): alkd = y(5)
RETURN
'*************************************************************************
```

```
PRINTER:    'Subroutine writes a file for subsequent plotting
GOSUB DEFINITIONS
GOSUB OTHER                         'print current values
PRINT #1, x; pco2; sigcs; sigcd; alks; alkd; hco3s; co3s
GOSUB PLOTTER
RETURN
'**********************************************************************
EQUATIONS: 'The differential equations that define the simulation
GOSUB DEFINITIONS
GOSUB OTHER
yp(1) = (pco2s - pco2) / distime
yp(2) = -(pco2s - pco2) / distime * matmco2 - (1 + corat) * prod
yp(2) = (yp(2) + (sigcd - sigcs) * wflux) / vols       'continuation
yp(3) = ((1 + corat) * prod - (sigcd - sigcs) * wflux) / vold
yp(4) = ((alkd - alks) * wflux - (2 * corat - .15) * prod) / vols
yp(5) = ((2 * corat - .15) * prod - (alkd - alks) * wflux) / vold
RETURN
'**********************************************************************
OTHER: 'Subroutine evaluates quantities that change with time but that
'are not dependent variables
'Carbonate equilibria in surface ocean
sigc = sigcs: alk = alks: watemp = watemps
GOSUB CARBONATE
hco3s = hco3: co3s = co3
pco2s = kco2 * hco3s ^ 2 / co3s
RETURN
'**********************************************************************
CARBONATE:      'Subroutine solves the carbonate equilibria
'Equilibrium constants from Broecker and Peng (1982, p. 151)
kcarb = .000575 + .000006 * (watemp - 278)     'temperature dependence
kco2 = .035 + .0019 * (watemp - 278)
hco3 = sigc - SQR(sigc ^ 2 - alk * (2 * sigc - alk) * (1 - 4 * kcarb))
hco3 = hco3 / (1 - 4 * kcarb)                   'continuation
co3 = (alk - hco3) / 2
RETURN
'**********************************************************************
SPECS: 'Subroutine to read in the specifications of the problem
wflux = .001     '10^18 m^3/y
vold = 1.23      '10^18 m^3
vols = .12
matmco2 = .0495  '10^18 mole.  1 PAL = 280 ppm
prod = .000175   '10^18 mole/y
```

```
corat = .25        'ratio of carbonate to organic carbon in particles
sigcs = 2.01       'mole/m3
sigcd = 2.2
alks = 2.2
alkd = 2.26
distime = 8.64     'years
watemps = 288      'temperature of surface water, deg K
pco2 = 1
y(1) = pco2
y(2) = sigcs
y(3) = sigcd
y(4) = alks
y(5) = alkd
RETURN
'****************************************************************************
GRAFINIT:                 'Initialize graphics
'Plot relative departures from PLOTZ
'Sensitivity of the plot depends on PLOTS
FOR jplot = 1 TO numplot
    plotz(jplot) = y(jplot)
    plots(jplot) = y(jplot) / 5
NEXT jplot
'The labels are PLOTL$
plotl$(1) = "pco2": plotl$(2) = "sgcs": plotl$(3) = "sgcd"
plotl$(4) = "alks": plotT$(5) = "alkd"
GOSUB GRINC
RETURN
END
'****************************************************************************
GRINC:      'This is the core of GRAFINIT.  Not specific to one problem
SCREEN 1
COLOR 1, 2
CLS
VIEW (45, 15)-(315, 180), 1, 2
LOCATE 3, 1
PRINT "TIME"
FOR jrow = 1 TO numplot
    LOCATE 3 + jrow, 1
    PRINT plotl$(jrow)
NEXT jrow
LOCATE 1, 1
PRINT "X =                 DELX =";
```

```
LOCATE 25, 1
PRINT "BIGONE = ";
LOCATE 25, 20
PRINT "COUNT = ";
yval = .25
FOR jcount = 1 TO 3              'Tick marks
    yval = yval * 4
    ord = 135 + yval / (1 + yval) * 135
    LINE (ord, 161)-(ord, 165), 3
    ord = 135 - yval / (1 + yval) * 135
    LINE (ord, 161)-(ord, 165), 3
NEXT jcount
LINE (135, 161)-(135, 165), 3
LOCATE 23, 30
PRINT "REL.Y  -4    -1      0       1    4 16";
RETURN
'*****************************************************************************
PLOTTER:        'Subroutine plots values as the calculation proceeds
'The scale is normalized relative to the starting values
FOR jplot = 1 TO numplot       'Which values to plot
    ploty(jplot) = y(jplot)
NEXT jplot
GOSUB PLTC
RETURN
END
'*****************************************************************************
PLTC:           'The core of PLOTTER.  Not specific to one problem
'Keeping track of the time
LINE (0, 0)-(270, 160), 1, BF                   'Clear the box
timefrac = (x - xstart) / (xend - xstart)       'Fraction completed
LINE (0, 0)-(timefrac * 270, 8), 0, BF
FOR jrow = 1 TO numplot
    depart = ploty(jrow) - plotz(jrow)        'Plot departures from PLOTZ
    ord = depart / (plots(jrow) + ABS(depart)) * 135 + 135 'Sensitivity
    LINE (135, jrow * 8)-(ord, (jrow + 1) * 8), 2, BF
NEXT jrow
LOCATE 1, 5
PRINT x;
LOCATE 1, 26
PRINT "
LOCATE 1, 26
PRINT delx;                          'Watching values of parameters
```

```
LOCATE 25, 10
PRINT bigone;
LOCATE 25, 30
PRINT count;
RETURN
'**********************************************************************
FILER:              'Subroutine to set up files for input and output
PRINT
FILES "*.prn"
PRINT
BEEP
PRINT "Hit ENTER if you want results in file RESULTS.PRN."
INPUT "Otherwise enter the name of the file for results: ", resf$
IF resf$ = "" THEN
    OPEN "RESULTS.PRN" FOR OUTPUT AS #1
ELSE
    OPEN resf$ FOR OUTPUT AS #1
END IF
PRINT
BEEP
PRINT "The final values are stored for a possible continuation of the"
PRINT "calculation or for use as initial conditions."
PRINT "Hit ENTER if you want the final values in file ENDVAL.PRN."
INPUT "Otherwise enter the name of the file for final values: ", endf$
IF endf$ = "" THEN endf$ = "ENDVAL.PRN"
PRINT
BEEP
PRINT "Type P and hit RETURN to use the values in the program"
PRINT "as initial conditions."
PRINT "Otherwise hit ENTER to start the calculation with initial"
PRINT "conditions from file STARTER.PRN"
PRINT "Otherwise enter the name of the file to use for initial"
INPUT "conditions: ", incf$
IF incf$ = "p" OR incf$ = "P" THEN
    RETURN
ELSEIF incf$ = "" THEN
    OPEN "STARTER.PRN" FOR INPUT AS #3
ELSE
    OPEN incf$ FOR INPUT AS #3
END IF
GOSUB STARTER
RETURN
```

```
'*********************************************************************
STARTER: 'Subroutine reads initial values from a file
FOR jrow = 1 TO nrow
    INPUT #3, y(jrow)
NEXT jrow
INPUT #3, x
CLOSE #3
RETURN
'*********************************************************************
STOPPER:  'This subroutine saves the final values for a possible
'continuation or for use as initial conditions in future calculations
OPEN endf$ FOR OUTPUT AS #2
FOR jrow = 1 TO nrow
    PRINT #2, y(jrow)
NEXT jrow
PRINT #2, x
RETURN
'*********************************************************************
Plus subroutines GAUSS and SWAPPER from Program DGC03
Subroutine STEPPER from Program DGC04
Subroutine CHECKSTEP from Program DGC06
Subroutine SLOPER from Program DGC08
```

state and then to use the final values of the dependent variables as the starting values in future calculations.

To make this whole process easy I have added to my program three subroutines, FILER, STARTER, and STOPPER. Subroutine FILER first lists the available files with the suffix .PRN. It then sends to the screen a message asking where the results should be stored. The default is RE-SULTS.PRN, but I can enter another file name if I do not want to erase the current contents of RESULTS.PRN. During the calculation, subroutine PRINTER will write results to the file chosen at this time.

Subroutine FILER then asks what starting values to use. A response of p or P (for program) causes the program to use the starting values specified in subroutine SPECS. This option can be used when suitable files of starting values have not yet been generated. Alternatively, I can enter the name of a file. The default is STARTER.PRN. If the choice is to read starting values from a file, the work will be done by subroutine STARTER.

Finally, subroutine FILER asks where it should store the values calculated on the last step. The default is ENDVAL.PRN. These values can be used as starting values in future calculations. Subroutine STOPPER writes the final values to the file chosen at this time before closing all files and sounding a tone to indicate that the calculation is finished.

5.4.2 Mnemonic Names for Variables

With systems of increasing complexity, it becomes more and more difficult to keep track of which elements of the Y and YP arrays correspond to which physical variables, and such confusion leads to errors. I find it helpful to assign names to the variables that remind me what they are and relate these names to the elements of the *y* array in the subroutine DEFINITIONS. I call DEFINITIONS whenever I need current values of the dependent variables, but I can write expressions like those for *yp* in subroutine EQUATIONS using names rather than numbers.

5.4.3 Other Quantities That Vary

In addition to the dependent variables specified by differential equations in subroutine EQUATIONS, there are related quantities specified by algebraic functions of time and the dependent variables. Examples in this program are the concentrations of individual dissolved carbon species. To keep EQUATIONS free of clutter, I calculate these quantities in subroutine OTHER, which I call at the beginning of EQUATIONS. Parameters that do not vary, including the starting values of the dependent variables, are specified in subroutine SPECS, called just once at the beginning of the calculation.

5.4.4 Subroutines and Global Variables

Different versions of BASIC have more or less capability with respect to subroutines. I believe the GOSUB and RETURN form that I am using is the most widely available. Its most important feature is both a convenience and a potential source of error. All subprograms have access to all variables, which is convenient because the programmer does not have to specify in a calling sequence the variables that the subroutine is to use, with the attendant worries about the completeness and correctness of the calling sequence. On the other hand, a name must not be used with different meanings in different subroutines, as such use can result in quite unexpected changes in the values of variables. For this reason, I generally use fairly long names with some mnemonic value; it is dangerous to use short unspecific names like *a, b,* or *c* because you are likely to use them again for something else in some other part of your program.

In order to use a single subroutine to perform the same calculation on different input values, I must first specify the values, then call the subroutine, and finally save the result. An example here is the call to CARBONATE in subroutine OTHER. I have set this call up with what looks in this program like needless complexity so that I can later make a second call to CARBONATE in order to calculate deep-water properties as well as shallow-water properties.

5.4.5 Watch the Calculation as It Happens

It usually takes some time to get a program working correctly, and while the program is being debugged and tuned, it is helpful to be able to watch the changes in the various parameters as a calculation proceeds. Because it is tedious to watch numbers and hard to follow more than one or two variables as numbers scroll by on the screen, a graphical display is needed. It should be compact, so that many variables can be observed at the same time without confusion. It should be general, so that scales and amplification factors do not need to be reset each time a program is modified. And above all, it should not cause the program to halt whenever a variable exceeds its expected limits.

In an effort to meet these needs, I have developed the subroutines PLOTTER and GRAFINIT. PLOTTER is the working routine, and GRAFINIT is called just once to set up the screen, labels, and axes. I use horizontal bars to represent the values of the various parameters to be watched, with the top bar showing the progress of the calculation in time. This bar starts at the left edge of the panel and moves to the right in proportion to simulation time; the calculation stops when the bar reaches the right edge of the panel. To accommodate a calculation covering a specified time period rather than a specified number of steps, I have modified subroutine CORE to repeat the calculation while x varies from *xstart* to *xend*. The parameter *count* keeps track of the number of successful steps so that I know how large a RESULTS file is being written. The time is given by x, and the current value of the time step is shown by *delx*. *bigone*, determined in subroutine CHECKSTEP, indicates which dependent variable is changing the most rapidly and therefore is restricting the value of *delx*. Because program errors frequently show up as unreasonably small values of *delx*, it is helpful to know which variable is the immediate cause of the problem (see Figure 5–2).

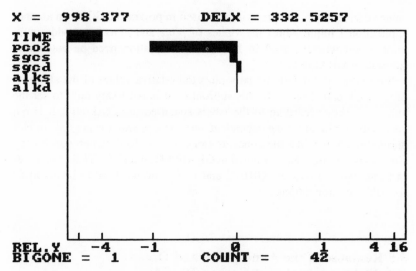

Fig. 5–2. Image of computer screen during a calculation with program DGC09.

In order to achieve flexibility, sensitivity, and robustness, I plot departures of the variables of interest from specified reference values. A bar projecting to the right of the middle of the panel indicates a value larger than the reference value. A bar projecting to the left indicates a smaller value. The length of the bar is proportional to

$$[ploty - plotz]/[plots + \text{ABS}(ploty - plotz)]$$

where *ploty* is the current value of the variable, *plotz* is the reference value, and *plots* controls the sensitivity, or how far the bar moves for a given change in the variable. This scaling law has the virtue of never going off screen: A very large departure from the reference value plots all the way to the right, and a very small value plots all the way to the left. PLOTTER is therefore robust, not causing the program to stop. The format of horizontal bars is compact, making it possible to follow as many as nineteen variables with the scales as I have written them. Sensitivity is easily adjusted to provide a responsive and informative display.

The parameters of the display are established in GRAFINIT, and the number of variables to watch is specified by *numplot*. There must be a corresponding number of reference values in the *plotz* array, sensitivity values in the *plots* array, and labels in the *plotl$* array. As a convenient start, I set *plotz* equal to the initial values and *plots* equal to the initial values divided by 5, the prescription I used in program DGC09. In another application I might adjust the values of *plotz* and *plots* to provide more information where I need it. No other information need be entered in subroutine GRAFINIT.

Subroutine PLOTTER actually plots the relative values of the elements of an array called *ploty*. In this subroutine it is necessary only to set the values of *ploty* according to the labels specified in GRAFINIT. It is not necessary to plot all of the dependent variables, *y*, and it is possible to plot derived quantities like the concentrations of dissolved carbon species, by simply specifying what is wanted in GRAFINIT and PLOTTER. The cores of these two subroutines, GRINC and PLTC, do not need to be modified for different simulations.

5.5 Response of the Atmosphere and Ocean to a Sudden Injection of Carbon Dioxide

Now that we have explained the various parts and procedures of program DGC09, we can put it to work. I find that the system requires some tuning to achieve steady-state values that correspond to the present-day observations summarized by Broecker and Peng (1982, p. 69). Tuning is needed because I am working with a few numbers to represent the average values of alkalinity and total carbon averaged over great volumes of the ocean. Similarly, I am using single values of the equilibrium constants to represent averages over seawater with variable temperatures. The numbers that I quoted earlier may not be precisely the right averages, and so to make a simple calculation behave, I tune it.

The difference between the total dissolved carbon in the surface and in deep-sea reservoirs depends on productivity. And the difference between the alkalinity in these reservoirs depends on productivity and also *corat*, the calcium-carbonate-to-organic-carbon ratio. The carbon dioxide partial pressure depends on the difference between total carbon and alkalinity in the surface reservoir, and all these depend on the total amount of carbon and alkalinity at the start of the calculation in the three reservoirs combined. By adjusting the values of these various parameters and repeating the calculation, I arrive at the following values for a steady-state system that is close to the present-day ocean with a preindustrial level of atmospheric carbon dioxide:

pco2	sigcs	sigcd	alks	alkd	hco3s	co3s
0.99801	2.02848	2.24723	2.19886	2.26011	1.8346	0.18212

For *prod* and *corat* I use the values listed in Section 5.3.

This program can now be used to calculate the response of the ocean and atmosphere to a sudden input of fossil fuel carbon dioxide. I simply change the starting value of *pco2* to 5 by inserting the statement

$$y(1) = 5$$

right after the call to subroutine FILER. The value of 5 replaces 0.99801 as the initial value for *pco2*, and the system is no longer in steady state. The results of the calculation are shown in Figure 5–3. The addition of carbon to the system has no effect on the alkalinity in either reservoir, and so I have not plotted the alkalinity values. There is a rapid initial decrease in *pco2* accompanied by an increase in *sigcs* as carbon is transferred from the atmosphere to shallow ocean. This is the response calculated previously,

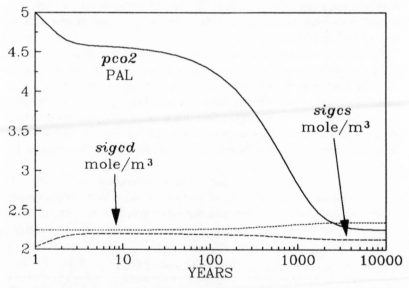

Fig. 5–3. The response of carbon in atmosphere and ocean to a sudden increase of atmospheric carbon dioxide to a value of 5 PAL. Alkalinity is not plotted because it does not change in response to this perturbation.

with the results plotted in Figure 2–3. But after a few years in the current calculation, the shallow ocean and atmosphere are balanced, and the rapid initial adjustment stops. On a much longer time scale, carbon is transferred from the shallow sea to the deep sea, leading to a further decrease in atmospheric carbon dioxide. The system ceases to evolve after a few thousand years, a few times the residence time of carbon in the deep sea. Both carbon and alkalinity are conserved in this system, and so the calculation simply redistributes a fixed amount of carbon among the three reservoirs. The accuracy of the calculation can be checked by converting from concentrations and pressure units to masses of carbon and summing.

5.6 How to Tabulate a History for Maximum Flexibility

On a human time scale, the input of fossil fuel carbon dioxide is not sudden. The rate is expected to increase for several hundred years and then to decrease again as reserves are exhausted, as shown in Table 5-1 (Broecker and Peng, 1982, p. 553). The program being developed here can be used to simulate the response of the ocean and atmosphere to a gradual input of carbon dioxide.

It is possible to find an algebraic expression that describes reasonably well an assumed evolution of the carbon dioxide source, but it is more generally useful to be able to work with a tabulated history. Using a tabulated history requires a subroutine that will look up values in the table and interpolate between entries. I find that a linear interpolation provides adequate accuracy for typical problems in global change. A smoothly varying function can always be reproduced by adding enough entries to the table.

In the following program, my table is called *table(nent, nvar)*, in which the first column is the independent variable, time, which must increase monotonically down the table, and the other columns are one or more tabulated variables. In this example there is only one dependent variable,

Table 5–1. Rate of Release of Fossil Fuel Carbon Dioxide

DATE (A.D.)											
1850	1950	1980	2000	2050	2080	2100	2120	2150	2225	2300	2500
10^{14} mole/y											
0	1	4	5	8	10	10.5	10	8	3.5	2	0

```
'Program DGC10 solves the carbon system in atmosphere, shallow,
'and deep sea.
'With a gradual injection of fossil fuel carbon dioxide
nrow = 5                'the number of equations and unknowns
ncol = nrow + 1
'==================================================================
DIM sleq(nrow, ncol), unk(nrow), y(nrow), dely(nrow), yp(nrow)
'Establish arrays for graphics routines, GRAFINIT and PLOTTER
numplot = 6             'Number of variables to plot. Not more than 19
DIM plotz(numplot), plots(numplot), plotl$(numplot), ploty(numplot)
nent = 12: nvar = 2    'parameters of TLU
DIM table(nent, nvar), interpol(nvar)
jpoint = 1
GOSUB FFUEL
'==================================================================
GOSUB SPECS
GOSUB FILER
xstart = 1000              'time to start
xend = 7000           'time to stop, in years
x = xstart
delx = 10
mstep = 500
GOSUB CORE
END
'*****************************************************************************
FFUEL:  'Fossil fuel source of carbon dioxide
'Broecker and Peng, 1982, p. 553.  Units 10^14 mole/y
DATA 1850,0
DATA 1950,1
DATA 1980,4
DATA 2000,5
DATA 2050,8
DATA 2080,10
DATA 2100,10.5
DATA 2120,10
DATA 2150,8
DATA 2225,3.5
DATA 2300,2
DATA 2500,0
FOR jent = 1 TO nent
    FOR jvar = 1 TO nvar
        READ table(jent, jvar)
```

```
      NEXT jvar
   NEXT jent
   FOR jent = 1 TO nent              'Convert to 10^18 mole/y
      table(jent, 2) = .0001 * table(jent, 2)
   NEXT jent
   RETURN
'*************************************************************************
TLU:  'Subroutine performs linear interpolation of terms in TABLE
'The time is TVAR (= X + DELX for the reverse Euler method)
   IF tvar > table(1, 1) THEN 810
   FOR jvar = 2 TO nvar                     'if TVAR is out of range
      interpol(jvar) = table(1, jvar) 'interpolated values
   NEXT jvar
   GOTO 850
810    IF tvar < table(nent, 1) THEN 820
   FOR jvar = 2 TO nvar                     'if TVAR is out of range
      interpol(jvar) = table(nent, jvar) 'interpolated values
   NEXT jvar
   GOTO 850
820    IF tvar > table(jpoint, 1) THEN 830      'start search at last find
   jpoint = jpoint - 1                       'step back if needed
   GOTO 820
830    IF tvar < table(jpoint + 1, 1) THEN 840 'TVAR is bracketed
   jpoint = jpoint + 1                       'step forward if needed
   GOTO 830
'If we get this far, we should have TVAR located on the table
840    dtvar = tvar - table(jpoint, 1)
      FOR jvar = 2 TO nvar
      slvar = table(jpoint + 1, jvar) - table(jpoint, jvar)'slope
      slvar = slvar / (table(jpoint + 1, 1) - table(jpoint, 1))
      interpol(jvar) = table(jpoint, jvar) + slvar * dtvar'interpolate
   NEXT jvar
850 RETURN
'*************************************************************************
DEFINITIONS:     'Subroutine defines dependent variables
pco2 = y(1): sigcs = y(2): sigcd = y(3): alks = y(4): alkd = y(5)
RETURN
'*************************************************************************
PRINTER:   'Subroutine writes a file for subsequent plotting
GOSUB DEFINITIONS
tvar = x                 'Time for interpolation
GOSUB OTHER                     'print current values
```

```
PRINT #1, x; pco2; sigcs; sigcd; alks; alkd; hco3s; co3s; fuel
GOSUB PLOTTER
RETURN
'**************************************************************************
EQUATIONS: 'The differential equations that define the simulation
GOSUB DEFINITIONS
tvar = x + delx          'Time for interpolation
GOSUB OTHER
yp(1) = (pco2s - pco2) / distime + fuel / matmco2
yp(2) = -(pco2s - pco2) / distime * matmco2 - (1 + corat) * prod
yp(2) = (yp(2) + (sigcd - sigcs) * wflux) / vols        'continuation
yp(3) = ((1 + corat) * prod - (sigcd - sigcs) * wflux) / vold
yp(4) = ((alkd - alks) * wflux - (2 * corat - .15) * prod) / vols
yp(5) = ((2 * corat - .15) * prod - (alkd - alks) * wflux) / vold
RETURN
'**************************************************************************
OTHER: 'Subroutine evaluates quantities that change with time but that
'are not dependent variables
'Carbonate equilibria in surface ocean
sigc = sigcs: alk = alks: watemp = watemps
GOSUB CARBONATE
hco3s = hco3: co3s = co3
pco2s = kco2 * hco3s ^ 2 / co3s
GOSUB TLU
fuel = interpol(2)
RETURN
'**************************************************************************
GRAFINIT:                 'Initialize graphics
'Plot relative departures from PLOTZ
'Sensitivity of the plot depends on PLOTS
FOR jplot = 1 TO nrow
    plotz(jplot) = y(jplot)
    plots(jplot) = y(jplot) / 5
NEXT jplot
plotz(6) = 0: plots(6) = .001: plotl$(6) = "fuel"
'The labels are PLOTL$
plotl$(1) = "pco2": plotl$(2) = "sgcs": plotl$(3) = "sgcd"
plotl$(4) = "alks":plotl$(5) = "alkd"
GOSUB GRINC
RETURN
END
'**************************************************************************
```

```
PLOTTER:        'Subroutine plots values as the calculation proceeds
'The scale is normalized relative to the starting values
FOR jplot = 1 TO nrow        'Which values to plot
    ploty(jplot) = y(jplot)
NEXT jplot
ploty(6) = fuel
GOSUB PLTC
RETURN
END
'******************************************************************************
Plus subroutines GAUSS and SWAPPER from Program DGC03
Subroutine STEPPER from Program DGC04
Subroutine CHECKSTEP from Program DGC06
Subroutine SLOPER from Program DGC08
Subroutines STOPPER, STARTER, FILER, PLTC, GRAFINIT, GRINC,
    SPECS, CARBONATE, CORE from Program DGC09
```

fuel. *nvar*, the number of columns in the table, is therefore 2. *nent* is the number of rows. The table is defined in program DGC10 by subroutine FFUEL, with the table lookup and interpolation performed by subroutine TLU and the interpolated values of the dependent variables returned in the array *interpol(nvar)*.

Using *tvar* for the time at which I want the interpolated values, I search through the table until I find the entries that bracket *tvar*. I then interpolate between these bracketing values. Because the calculation generally proceeds forward in time, I remember *jpoint*, the current location in the table, and restart the search at *jpoint* on successive calls to TLU. But because a reduction in *delx* causes the next call to the table to be for an earlier time, the search can go up the table as well as down. If *tvar* is less than the smallest time in the table, I set the dependent variables equal to the first entries in the table. If *tvar* is greater than the largest time in the table, I set the dependent variables equal to the last entries.

As a precautionary note, remember that *tvar* = *x* + *delx* in the application of the reverse Euler method of solution but that *tvar* = *x* when the dependent variable is being evaluated for printing or plotting.

The results of the calculation over 800 years are shown in Figure 5–4. The horizontal axis here is the date: Today we are still in the early stages of the predicted increase in atmospheric carbon dioxide. The peak is expected to occur about A.D. 2350, when there will be about four times the current

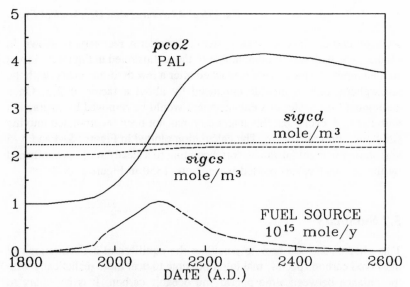

Fig. 5–4. The response of carbon in atmosphere and ocean to a fossil fuel source of carbon dioxide that increases for a few hundred years and then decreases. The history of the source is plotted at the bottom of the figure (Broecker and Peng, 1982, p. 553).

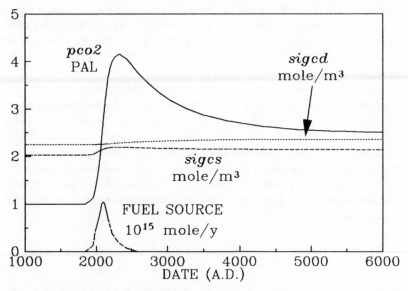

Fig. 5–5. The long-term response of carbon in atmosphere and ocean to a fossil fuel source distributed over hundreds of years.

level of carbon dioxide. The system's long-term recovery is shown in Figure 5–5, with results comparable to those illustrated in Figure 5–3 for a sudden input. A steady state is reached after a few thousand years, with the atmospheric carbon dioxide enhanced by about a factor of 2.5. On a geological time scale, this enhancement would be removed by interaction with the rock cycle, but this interaction has not been incorporated into the simulation presented here. The calculation plotted in Figures 5–4 and 5–5 was started with initial values corresponding to a steady state, as confirmed by the constant values plotted before A.D. 1850 in Figure 5–5.

5.7 Summary

This chapter showed how to calculate the equilibrium concentrations of dissolved carbon species, making it possible to deal more realistically with the balance between atmospheric and oceanic carbon. It is necessary to keep track of both total dissolved carbon and alkalinity. The subroutine that calculates the species concentrations is CARBONATE, introduced in program DGC09. Program DGC10 simulated the response of ocean and atmosphere to the gradual release of fossil fuel carbon dioxide over a period of centuries. For this simulation I introduced the capability of reading in a prescribed history in the form of a table and then interpolating in this table to obtain values of the specified quantity, in this case the rate of release of carbon dioxide, at whatever times are called for by the calculation. The interpolation routine is subroutine TLU.

Much of the chapter was devoted to introducing subroutines and procedures that make it easier to deal with a variety of simulations without writing much new code. The most important of these, in addition to TLU, are FILER, STOPPER, and STARTER, which manage files, and PLOTTER and GRAFINIT, which handle a graphic presentation of results during a calculation. I also called attention to the need to use unique names for different variables in different subroutines, suggested the use of mnemonic variable names, and discussed tuning and the achievement of steady-state initial values.

6 How to Calculate Isotope Ratios

6.1 Introduction

The calculation of isotope ratios requires special consideration because isotope ratios, unlike matter or energy, are not conserved. In this chapter I shall show how extra terms arise in the equations for the rates of change of isotope ratios. The equations developed here are quite general and can be applied to most of the isotope systems used in geochemistry. As an example of the application of these new equations, I shall demonstrate a simulation of the carbon isotopic composition of ocean and atmosphere and then use this simulation to examine the influence on carbon isotopes of the combustion of fossil fuels. As an alternative application I shall simulate the carbon isotopic composition of the water in an evaporating lagoon and show how the composition and other properties of this water might be affected by seasonal changes in evaporation rate, water temperature, and biological productivity.

6.2 How to Calculate Isotope Ratios

Equations for the rates of change of individual isotopes in a reservoir are not essentially different from the equations for the rates of change of chemical species. Isotopic abundances, however, are generally expressed as ratios of one isotope to another and, moreover, not just as the ratio but also as the departure of the ratio from a standard. This circumstance introduces some algebra into the derivation of an isotopic conservation equation. It is convenient to pursue this algebra just once, as I shall in this section, after which all isotope simulations can be formulated in the same way. I shall use the carbon isotopes to illustrate this derivation, but the same approach can be used for the isotopes of other elements, such as sulfur, oxygen, nitrogen, hydrogen, or strontium.

The most abundant isotope of carbon has a mass of 12 atomic mass units, ^{12}C. A less abundant stable isotope is ^{13}C. And much less abundant is the radioactive isotope ^{14}C, also called *radiocarbon*. It is convenient to express the abundances of these rare isotopes in terms of ratios of the number of atoms of the rare isotope in a sample to the number of atoms of the abundant isotope. We call this ratio r, generally a very small number. To arrive at numbers of convenient magnitude, it is conventional to express the ratio in terms of the departure of r from the ratio in a standard, which I call s, and to express this departure in parts per thousand of s. Thus the standard delta notation is

$$\delta^{13}C \text{ or } ^{14}C = (r - s)/s * 1000 \text{ per mil}$$

This equation can be rearranged to derive an expression for the ratio

$$r = s * (1 + del/1000)$$

where *del* denotes δ^{13}C or ^{14}C.

Suppose that a given reservoir contains m atoms of the abundant carbon isotope and has an isotopic composition of *del*. Suppose that carbon of composition *deli* is supplied to the reservoir at rate *fi* while carbon of composition *delo* is removed from the reservoir at rate *fo*. The equation for the rate of change of the number of atoms of the abundant isotope in the reservoir is

$$\frac{dm}{dt} = fi - fo$$

The number of atoms of the rare isotope in the reservoir is $r * m$, and the rate of change of this number is

$$\frac{d(r * m)}{dt} = fi * ri - fo * ro$$

where $ri = s * (1 + deli/1000)$ and $ro = s * (1 + delo/1000)$. This derivation applies to a stable isotope. I shall consider the impact of radioactive decay later.

Expand the left-hand side and substitute for *ri* and *ro* on the right-hand side to obtain

$$r * \frac{dm}{dt} + m * \frac{dr}{dt} = s * (fi - fo) + (s/1000) * (fi * deli - fo * delo)$$

Move the first term from the left to the right, and substitute

$$r = s * (1 + del/1000) \text{ and } fi - fo = \frac{dm}{dt}$$

The result is

$$m * \frac{dr}{dt} = (s/1000) * (fi * deli - fo * delo - del * \frac{dm}{dt})$$

From the expression for r in terms of *del*,

$$\frac{dr}{dt} = (s/1000) * \frac{ddel}{dt}$$

so

$$\frac{ddel}{dt} = (fi * deli - fo * delo - del * \frac{dm}{dt})/m$$

This is the expression I use to calculate isotope ratios. The rate of change of delta is the influx of the abundant isotope times the delta of the source minus the outflux times the delta of the sink minus an extra term, delta times the rate of change of the amount of the abundant isotope ($= fi - fo$), with the whole expression divided by the amount m of the abundant isotope. The isotope ratio in the standard, s, does not enter the expression, nor do the units, whether per mil or percent.

I turn now to the expression for a radioactive isotope, for example, radiocarbon. If *lambda* is the decay constant, the rate of loss of radioactive atoms by decay will be $lambda * r * m$. For the sake of generality, suppose that there is a source that generates radiocarbon at a rate equal to *qrc*. The conservation equation becomes

$$\frac{d(r * m)}{dt} = fi * ri - fo * ro - lambda * r * m + qrc$$

Substituting and rearranging as before,

$$m * \frac{dr}{dt} = (s/1000) * (fi * deli - fo * delo - del * \frac{dm}{dt})$$
$$- lambda * r * m + qrc$$

which leads to

$$\frac{ddel}{dt} = (fi * deli - fo * delo - del * \frac{dm}{dt})/m$$
$$- lambda * r/s * 1000 + qrc/m * 1000/s$$

But $r/s * 1000 = 1000 + del$, so the conservation equation in final form is

$$\frac{d del}{dt} = (fi * deli - fo * delo - del * \frac{dm}{dt})/m$$
$$- lambda * (1000 + del) + qrc/m * 1000/s$$

In this case the units do matter, but the standard ratio, s, enters only in the term that represents the production of radiocarbon.

6.3 Carbon Isotopes in the Ocean and Atmosphere

The application of these equations to the marine carbon system is illustrated by program ISOT01, which adds isotopes to the three-reservoir system of atmosphere, shallow ocean, and deep ocean presented in program DGC10 in Chapter 5. In subroutine EQUATIONS, equations 6 to 8 are for the stable isotope ^{13}C, and equations 9 to 11 describe radiocarbon. The rest of the physical system is identical to that of program DGC10.

Isotopic fractionation occurs during photosynthesis, which produces organic carbon deficient in ^{13}C by an amount *delcorg* compared with the dissolved carbon in the surface-water reservoir where photosynthesis occurs. The corresponding term in equations 7 and 8 is *prod* * (*dcs* − *delcorg*). The descent of isotopically light particulate carbon into the deep sea leaves the surface ocean isotopically heavy, and the decay of the particles in the deep sea makes deep-sea carbon isotopically light. The formation of particles of calcium carbonate particles does not fractionate isotopes, and so the corresponding term is *prod* * *corat* * *dcs*. There is isotopic fractionation between gaseous carbon dioxide and the dissolved carbon in the surface ocean. The amount is *delcatm*, which depends weakly on temperature (Broecker and Peng, 1982, p. 310). The value of *delcatm* is calculated in subroutine OTHER. I have assumed that the fractionation between gas and dissolved carbon occurs entirely during the flow of carbon from the ocean to the atmosphere, although some part of the total fractionation may in fact occur on the return flow (Siegenthaler and Münnich, 1981). The terms in equations 6 and 7 are *pco2s* * (*dcs* − *delcatm*) − *pco2* * *dcp*. The fractionation factors for radiocarbon in photosynthesis and in the transfer from ocean to atmosphere are just twice as large as the factors for ^{13}C because the difference in masses is twice as large. This is the reason for the factors of 2 multiplying *delcorg* and *delcatm* in the equations for radiocarbon.

```
'Program ISOT01 solves the carbon system in atmosphere, shallow,
'and deep sea.  Includes a gradual injection of fossil fuel carbon
'dioxide, read from a file.
nrow = 11        'the number of equations and unknowns
ncol = nrow + 1
'===================================================================
DIM sleq(nrow, ncol), unk(nrow), y(nrow), dely(nrow), yp(nrow)
DIM incind(nrow)         'Controls test on increment in CHECKSTEP
'Establish arrays for graphics routines, GRAFINIT and PLOTTER
numplot = 12        'Number of variables to plot.  Not more than 19
DIM plotz(numplot), plots(numplot), plotl$(numplot), ploty(numplot)
nent = 134: nvar = 2   'parameters of TLU
DIM table(nent, nvar), interpol(nvar)
jpoint = 1
'===================================================================
GOSUB SPECS
GOSUB FFUEL
GOSUB FILER
xstart = 1800        'time to start
xend = 1990        'time to stop, in years
x = xstart
delx = 1
mstep = 1
GOSUB CORE
END
'*******************************************************************
DEFINITIONS: 'Subroutine defines dependent variables
pco2 = y(1): sigcs = y(2): sigcd = y(3): alks = y(4): alkd = y(5)
dcp = y(6): dcs = y(7): dcd = y(8)
drp = y(9): drs = y(10): drd = y(11)
RETURN
'*******************************************************************
PRINTER:  'Subroutine writes a file for subsequent plotting
GOSUB DEFINITIONS
tvar = x        'Time for interpolation
GOSUB OTHER            'print current values
PRINT #1, x; pco2; sigcs; sigcd; alks; alkd; hco3s; co3s; fuel;
PRINT #1, dcp; dcs; dcd; drp; drs; drd
GOSUB PLOTTER
RETURN
'*******************************************************************
EQUATIONS: 'The differential equations that define the simulation
```

```
GOSUB DEFINITIONS
tvar = x + delx      'Time for interpolation
GOSUB OTHER
yp(1) = (pco2s - pco2) / distime + fuel / matmco2
yp(2) = -(pco2s - pco2) / distime * matmco2 - (1 + corat) * prod
yp(2) = (yp(2) + (sigcd - sigcs) * wflux) / vols      'continuation
yp(3) = ((1 + corat) * prod - (sigcd - sigcs) * wflux) / vold
yp(4) = ((alkd - alks) * wflux - (2 * corat - .15) * prod) / vols
yp(5) = ((2 * corat - .15) * prod - (alkd - alks) * wflux) / vold
 'Equations for carbon 13
yp(6) = (pco2s * (dcs - delcatm) - pco2 * dcp) / distime
yp(6) = (yp(6) + fuel / matmco2 * dcfuel - dcp * yp(1)) / pco2
yp(7) = (pco2 * dcp - pco2s * (dcs - delcatm)) / distime * matmco2
yp(7) = yp(7) - prod * (1 * (dcs - delcorg) + corat * dcs)
yp(7) = (yp(7) + (sigcd * dcd - sigcs * dcs) * wflux) / vols
yp(7) = (yp(7) - dcs * yp(2)) / sigcs
yp(8) = prod * (1 * (dcs - delcorg) + corat * dcs)
yp(8) = (yp(8) - (sigcd * dcd - sigcs * dcs) * wflux) / vold
yp(8) = (yp(8) - dcd * yp(3)) / sigcd
 'Equations for radiocarbon
yp(9) = (pco2s * (drs - 2 * delcatm) - pco2 * drp) / distime
yp(9) = yp(9) + fuel / matmco2 * drfuel
yp(9) = (yp(9) - drp * yp(1) + drsource) / pco2 - lambda * (1000 + drp)
yp(10) = (pco2 * drp - pco2s * (drs - 2 * delcatm)) / distime * matmco2
yp(10) = yp(10) - prod * '(1 * (drs - 2 * delcorg) + corat * drs)
yp(10) = (yp(10) + (sigcd * drd - sigcs * drs) * wflux) / vols
yp(10) = (yp(10) - drs * yp(2)) / sigcs - lambda * (1000 + drs)
yp(11) = prod * (1 * (drs - 2 * delcorg) + corat * drs)
yp(11) = (yp(11) - (sigcd * drd - sigcs * drs) * wflux) / vold
yp(11) = (yp(11) - drd * yp(3)) / sigcd - lambda * (1000 + drd)
RETURN
'**************************************************************************
OTHER: 'Subroutine evaluates quantities that change with time but that
'are not dependent variables
'Carbonate equilibria in surface ocean
sigc = sigcs: alk = alks: watemp = watemps
GOSUB CARBONATE
hco3s = hco3: co3s = co3
pco2s = kco2 * hco3s ^ 2 / co3s
 'Isotope fractionation.  Broecker and Peng (1982, p. 310)
delcatm = 10.6 - (watemp - 273) / 10
GOSUB TLU
```

```
fuel = interpol(2)
RETURN
'***************************************************************************
SPECS: 'Subroutine to read in the specifications of the problem
wflux = .001    '10^18 m^3/y
vold = 1.23     '10^18 m^3
vols = .12
matmco2 = .0495 '10^18 mole.  1 PAL = 280 ppm
prod = .000175  '10^18 mole/y
corat = .25     'ratio of carbonate to organic carbon in particles
sigcs = 2.02848 'mole/m3
sigcd = 2.24723
alks = 2.19886
alkd = 2.26011
distime = 8.64 'years
watemps = 288  'temperature of surface water, deg K
pco2 = .99801
delcorg = 20        'Fractionation by phytoplankton
dcd = .3            'Initial isotope ratio for deep water
lambda = 1 / 8200    'Radiocarbon decay constant per year
drsource = 6.4       'Effective atmospheric source of radiocarbon
dcfuel = -25        'Carbon 13 in fossil fuel
drfuel = -1000       'Radiocarbon in fossil fuel
y(1) = pco2
y(2) = sigcs
y(3) = sigcd
y(4) = alks
y(5) = alkd
y(6) = dcd - 5.5
y(7) = dcd + 2.5
y(8) = dcd
y(9) = 0
y(10) = -50
y(11) = -20
FOR jrow = 1 TO 5    'Test relative increments
    incind(jrow) = 1
NEXT jrow
FOR jrow = 6 TO 8    'Test absolute increments for isotopes
    incind(jrow) = .2
NEXT jrow
FOR jrow = 9 TO 11   'Test absolute increments for isotopes
    incind(jrow) = 10
```

```
NEXT jrow
RETURN
'****************************************************************************
CHECKSTEP:   'This subroutine identifies the largest relative increment
'ABS(dely/y); compares it with the specified value MAXINC, and adjusts
'the step size if adjustment is needed.
'If INCIND<>1 then absolute increment is used.
'In this case the largest permitted increment is INCIND.
'If INCIND=0 the variable is not tested.
biginc = 0
FOR jrow = 1 TO nrow
'Find the largest relative increment or absolute increment,
'depending on INCIND
    IF incind(jrow) = 0 THEN
        relinc = 0
    ELSEIF incind(jrow) = 1 AND y(jrow) <> 0 THEN
        relinc = ABS(dely(jrow) / y(jrow))
    ELSE
        relinc = ABS(dely(jrow) / incind(jrow)) * maxinc
    END IF
    IF biginc < relinc THEN biginc = relinc
    IF biginc = relinc THEN bigone = jrow
NEXT jrow
IF biginc < maxinc THEN 1210
delx = delx / 2                   'reduce DELX and return without
RETURN                      'stepping forward
1210 GOSUB STEPPER               'increment Y and X and write file
IF biginc > maxinc / 2 THEN RETURN   'no change needed
delx = delx * 1.5           'increase DELX
IF delx > mstep THEN delx = mstep   'upper limit on DELX
RETURN
'****************************************************************************
GRAFINIT:          'Initialize graphics
'Plot relative departures from PLOTZ
'Sensitivity of the plot depends on PLOTS
plotz(1) = 1: plots(1) = .2
FOR jplot = 2 TO 5
    plotz(jplot) = y(jplot)
    plots(jplot) = y(jplot) / 5
NEXT jplot
plotz(12) = 0: plots(12) = .001: plotl$(12) = "fuel"
FOR jplot = 6 TO 8
```

```
    plotz(jplot) = 0        'Stable carbon isotopes
    plots(jplot) = 1
NEXT jplot
FOR jplot = 9 TO 11
    plotz(jplot) = 0        'Radioactive carbon isotopes
    plots(jplot) = 100
NEXT jplot
'The labels are PLOTL$
plotl$(1) = "pco2": plotl$(2) = "sgcs": plotl$(3) = "sgcd"
plotl$(4) = "alks": plotl$(5) = "alkd"
plotl$(6) = "dcp": plotl$(7) = "dcs": plotl$(8) = "dcd"
plotl$(9) = "drp": plotl$(10) = "drs": plotl$(11) = "drd"
GOSUB GRINC
RETURN
END
'****************************************************************************
PLOTTER:    'Subroutine plots values as the calculation proceeds
'The scale is normalized relative to the starting values
FOR jplot = 1 TO nrow    'Which values to plot
    ploty(jplot) = y(jplot)
NEXT jplot
ploty(12) = fuel
GOSUB PLTC
RETURN
END
'****************************************************************************
FFUEL: 'Fossil fuel source of carbon dioxide
OPEN "FOSSFUEL.PRN" FOR INPUT AS #4
'Data of Keeling and Rotty
FOR jent = 1 TO nent
    FOR jvar = 1 TO nvar
        INPUT #4, table(jent, jvar)
    NEXT jvar
NEXT jent
FOR jent = 1 TO nent        'Convert to 10^18 mole/y
    table(jent, 2) = .0001 * table(jent, 2)
NEXT jent
CLOSE #4
RETURN
'****************************************************************************
Plus subroutines GAUSS and SWAPPER from Program DGC03
Subroutine STEPPER from Program DGC04
```

```
Subroutine SLOPER from Program DGC08
Subroutine STARTER, STOPPER, FILER, GRINC, PLTC, CARBONATE,
   and CORE from Program DGC09
Subroutine TLU from Program DGC10
```

For radiocarbon, the standard ratio s is provided by the preindustrial atmosphere, for which $\delta = 0$. Cosmic rays interacting with atmospheric nitrogen were the main source of preindustrial radiocarbon. In the steady state, this source *drsource* is just large enough to generate an atmospheric delta value equal to zero. The source appears in equation 9 for atmospheric radiocarbon. Its value, specified in subroutine SPECS, I adjust to yield a steady-state atmospheric delta value of 0. The source balances the decay of radiocarbon in the atmosphere and in all of the oceanic reservoirs. Because radiocarbon has an overall source and sink—unlike the phosphorus, total carbon, ^{13}C, and alkalinity in this simulation—the steady-state values of radiocarbon do not depend on the initial values.

Fossil fuel source terms appear not only in equation 1 for atmospheric carbon dioxide *fuel/matmco2* but also in equations 6 and 9 for carbon isotopes in the atmosphere *fuel/matmco2 * dcfuel* or *drfuel*. The ^{13}C delta value for the fossil fuel source is *dcfuel* $= -25$, and the radiocarbon value is *drfuel* $= -1000$, because fossil carbon is devoid of radiocarbon, $r/s = 0$, and *del* $= -1 * 1000$.

In program DGC10 I used subroutine TLU to provide the history of fossil fuel release by linear interpolation in time between tabulated values. The tabulated values of fossil fuel release rate were read in subroutine FFUEL; for this first use of TLU I did not want to have too much detail. In fact, fossil fuel release rates have varied from year to year, so it is appropriate to work with a much finer time resolution. I achieve this in program ISOT01 with a completely new FFUEL subroutine, which reads the table of release rates at intervals of one year from a file. This file incorporates values published by Keeling (1973) and by Rotty and Marland (1986). The table has 134 rows and 2 columns, specified among the DIM statements at the beginning of the program. In order to keep the time steps of the solution commensurate with the detail in the source, I set the maximum step, *mstep*, to 1 year. As usual, I run the program without any fossil fuel input (from year $-100,000$ to $-10,000$) to achieve a steady state, adjusting the parameters to provide initial values that look reasonable. The starting values for the carbon system, total dissolved carbon, atmospheric pressure, and alkalinity are the steady-state values calculated with program DGC09. These values are specified in SPECS.

The study of isotopes makes it necessary to introduce a further refinement in the general method of solution. I have been using a test of the relative increment to adjust the time step. The *relative increment* is the change in a dependent variable divided by the value of that variable. This is not a useful test, however, when the value of the variable approaches zero, because the test requires progressively smaller time steps. None of the variables I considered in previous chapters has approached zero, and so there has been no problem with this test. But carbon isotope ratios of seawater have delta values near zero, and a problem may occur when calculating these values. I have modified subroutine CHECKSTEP to permit a flexible response to this situation.

Subroutine CHECKSTEP refers to an array *incind(nrow)* before performing its tests. Variables for which the corresponding element of *incind* is 1 are tested for relative increment in the manner of the earlier programs. Elements for which *incind* = 0 are excluded from the test altogether. They do not affect step size. The variables for which *incind* is neither 0 nor 1 are tested on the absolute increment. The step size will be reduced if the absolute increment for such variables exceeds the value of *incind*. The values of *incind* are set at the end of subroutine SPECS. I test relative increments for dissolved carbon, alkalinity, and atmospheric pressure as before, but absolute increments for the isotopes, requiring that the increments in $\delta^{13}C$ be less than 0.2 per mil and the increments in $\delta^{14}C$ be less than 10 per mil. Delta values for radiocarbon are much larger than for the stable isotope.

I have made some changes in GRAFINIT to provide more generally informative real-time plots. The reference value for atmospheric carbon dioxide is set to 1 PAL, so that with a sensitivity of 0.2, I look at departures from 1. As before, the reference values for alkalinity and dissolved carbon are the starting values. For the isotopes I plot departures of delta from 0, with a sensitivity of 1 for ^{13}C and 100 for ^{14}C.

6.4 The Influence of Fossil Fuel Combustion on Carbon Isotopes

The calculation is started with steady-state initial values in A.D. 1800 and carried forward by 1-year time steps to 1990. The results of the calculation are plotted in Figures 6–1, 6–2, and 6–3. The solid line at the bottom of

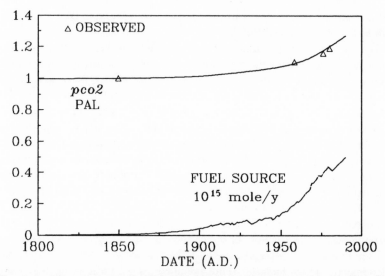

Fig. 6–1. The response to fossil fuel burning of atmospheric carbon dioxide. The fossil fuel combustion rate is shown at the bottom of the figure, and the calculated carbon dioxide level appears as a solid line at the top of the figure. Observed carbon dioxide values are plotted as triangles (Broecker and Peng, 1982). The observations have been normalized to preindustrial theoretical values.

Figure 6–1 shows the fossil fuel combustion rate, and the solid line at the top shows the calculated response of atmospheric carbon dioxide. The triangles in the figure give a few values of observed carbon dioxide partial pressures. Reference to Figure 5–4 shows that dissolved carbon concentrations change very little before the year 2000, and I therefore do not plot them. The results for these concentrations calculated with the new program are virtually identical to the results calculated with program DGC10.

Delta ^{13}C values in the atmosphere and ocean are shown in Figure 6–2. As might be expected, the atmosphere shows a pronounced decrease in isotope ratio because it is receiving isotopically light fossil fuel carbon dioxide. The response of the surface ocean on this time scale is somewhat smaller, reflecting the time it takes for the surface ocean to come to equilibrium with the atmosphere. It takes much longer for the added carbon to penetrate into the deep ocean, and so the results plotted in Figure 6–2 show essentially no change in deep-ocean carbon isotopes. This figure also plots observations taken from Broecker and Peng (1982). The observations are normalized to the calculated preindustrial isotope ratios. Only the depar-

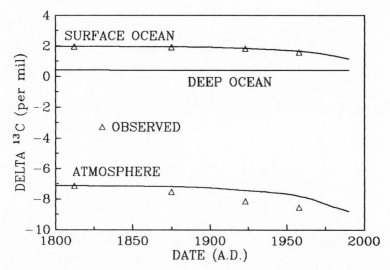

Fig. 6–2. The response to fossil fuel combustion of the ^{13}C ratio in the atmosphere, surface ocean, and deep ocean. Triangles show observed values from Broecker and Peng (1982). These observations have been normalized to preindustrial theoretical values.

tures from preindustrial values are significant, not the absolute levels. Although the simulation is behaving qualitatively as the real world does, the agreement between calculations and observations is not particularly good. This well-documented situation merits and has received elsewhere theoretical treatment with a more elaborate model of ocean circulation. But the purpose of our calculation is to show how isotope ratios can be calculated, not to provide a realistic simulation of a well-studied system.

The results for ^{14}C are plotted in Figure 6–3. Again, the response of the atmosphere is quite pronounced. The response of the shallow ocean is less marked, and the deep ocean shows no response at all on this time scale. Radiocarbon ratios are lower in the ocean than in the atmosphere because radioactive decay reduces the ^{14}C ratio. The difference between the steady-state atmosphere and the steady-state values in the oceanic reservoirs is an indication of how much time has elapsed since these masses of water last equilibrated with the atmosphere. Measurements of radiocarbon are an important source of information on the circulation of the deep ocean, and the differences between ^{13}C ratios in the different reservoirs have quite different causes: The deep ocean is lighter than the surface ocean because

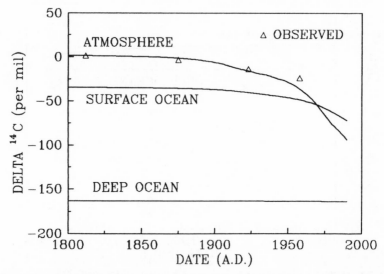

Fig. 6–3. The response to fossil fuel combustion of ^{14}C ratios in the atmosphere, shallow ocean, and deep ocean. Triangles show observations reported by Broecker and Peng (1982). The observations have been normalized to preindustrial theoretical values.

particulate carbon of biological origin is isotopically light. As it settles into the deep sea and is oxidized there, it drives down the isotope ratio of the deep sea and drives up the isotope ratio of the shallow sea. The difference between the shallow ocean and the atmosphere is a consequence of the fractionation between gaseous carbon dioxide and dissolved carbon dioxide.

6.5 Carbon Isotopes in an Evaporating Lagoon

The response of carbon isotopes in the atmosphere and ocean to fossil fuel burning is so familiar that the results of the calculation just carried out are not very interesting. It is, however, easy to adapt the program just discussed for application to a system for which the results are not immediately obvious. I will carry out an application of this kind in this section, just to have some fun with these equations and routines.

The system I shall simulate is an evaporating lagoon connected to the open ocean. The quantities to be calculated, listed in order in subroutine

```
'Program ISOTO2 solves the carbon system in an evaporating lagoon
nrow = 5        'the number of equations and unknowns
ncol = nrow + 1
'===================================================================
DIM sleq(nrow, ncol), unk(nrow), y(nrow), dely(nrow), yp(nrow)
DIM incind(nrow)        'Controls test on increment in CHECKSTEP
'Establish arrays for graphics routines, GRAFINIT and PLOTTER
numplot = 6     'Number of variables to plot. Not more than 19
DIM plotz(numplot), plots(numplot), plotl$(numplot), ploty(numplot)
'===================================================================
GOSUB SPECS
GOSUB FILER
xstart = x'1        'time to start
xend = xstart + 10      'time to stop, in years
x = xstart
delx = .01
mstep = .1
GOSUB CORE
END
'****************************************************************************
DEFINITIONS: 'Subroutine defines dependent variables
cas = y(1): sigcs = y(2): alks = y(3): dcs = y(4): drs = y(5)
RETURN
'****************************************************************************
PRINTER:  'Subroutine writes a file for subsequent plotting
GOSUB DEFINITIONS
tvar = x          'Time for seasonal change
GOSUB OTHER            'print current values
PRINT #1, x; cas; sigcs; alks; hco3s; co3s; fuel; dcs; drs; omegadel;
PRINT #1, -diss; pco2s
GOSUB PLOTTER
RETURN
'****************************************************************************
EQUATIONS: 'The differential equations that define the simulation
GOSUB DEFINITIONS
tvar = x + delx      'Time for seasonal change
GOSUB OTHER
yp(1) = (casw * fluxin - cas * fluxout) / depth + diss
yp(2) = -(pco2s - pco2) / distime * matcol / depth - prod + diss
yp(2) = yp(2) + (sigcsw * fluxin - sigcs * fluxout) / depth
yp(3) = 2 * diss + (alksw * fluxin - alks * fluxout) / depth
yp(4) = (pco2 * dcp - pco2s * (dcs - delcatm)) / distime * matcol / depth
```

```
yp(4) = yp(4) - prod * (dcs - delcorg) + diss * dcs
yp(4) = yp(4) + (sigcsw * dcsw * fluxin - sigcs * dcs * fluxout) / depth
yp(4) = (yp(4) - dcs * yp(2)) / sigcs
yp(5) = (pco2 * drp - pco2s * (drs - 2 * delcatm)) / distime
yp(5) = yp(5) * matcol / depth
yp(5) = yp(5) - prod * (drs - 2 * delcorg) + diss * drs
yp(5) = yp(5) + (sigcsw * drsw * fluxin - sigcs * drs * fluxout) / depth
yp(5) = (yp(5) - drs * yp(2)) / sigcs - lambda * (1000 + drs)
RETURN
'****************************************************************************
OTHER: 'Subroutine evaluates quantities that change with time but that
'are not dependent variables
fuel = SIN(6.242 * tvar)              'General purpose seasonal change
fluxin = 4 + fuel
'prod = fuel
'Carbonate equilibria in lagoon
sigc = sigcs: alk = alks: watemp = watemps ' + 5 * fuel
GOSUB CARBONATE
hco3s = hco3: co3s = co3
pco2s = kco2 * hco3s ^ 2 / co3s
omegadel = co3s * cas / csat - 1
'carbonate dissolution rate, mole/m^3/y
diss = (EXP(-disscon * omegadel) - EXP(pcpcon * omegadel)) * disfac
'Isotope fractionation.  Broecker and Peng (1982, p.310)
delcatm = 10.6 - (watemp - 273) / 10
RETURN
'****************************************************************************
SPECS: 'Subroutine to read in the specifications of the problem
fluxin = 4              'm/y
fluxout = 2             'm/y
depth = 5              'water depth in m
matmco2 = .0495 '10^18 mole.  1 PAL = 280 ppm
matcol = matmco2 * 10000! / 5.1
'CO2 in a column of atmosphere, mole/m^2
prod = 0              ' mole/m^3/y
sigcsw = 2!     'mole/m3
alksw = 2.2
casw = 10            'mole/m^3
distime = 8.64 'years
watemps = 288 'temperature of surface water, deg K
pco2 = 1!              'PAL
csat = .46  'calcite saturation constant in mole^2/m^6
```

```
'Broecker and Peng, p. 59
disscon = 7     'dissolution constant in DISS
pcpcon = 1   'carbonate precipitation constant
disfac = .01  'scaling factor in dissolution rate mole/m^3/y
delcorg = 10       'Fractionation by photosynthetic organisms
dcsw = 2           'Delta 13C isotope ratio for sea water, per mil
drsw = -50          'Delta 14C isotope ratio for sea water
dcp = -7           'Delta 13C isotope ratio for air, per mil
drp = 0         'Delta 14C isotope ratio for air
lambda = 1 / 8200      'Radiocarbon decay constant per year
y(1) = casw
y(2) = sigcsw
y(3) = alksw
y(4) = dcsw
y(5) = drsw
FOR jrow = 1 TO 3      'Test relative increments
    incind(jrow) = 1
NEXT jrow
incind(4) = .2           'Test absolute increments for isotopes
incind(5) = 10
RETURN
'****************************************************************************
GRAFINIT:          'Initialize graphics
'Plot relative departures from PLOTZ
'Sensitivity of the plot depends on PLOTS
FOR jplot = 1 TO 3
    plotz(jplot) = y(jplot)
    plots(jplot) = y(jplot) / 5
NEXT jplot
plotz(6) = 0: plots(6) = .2
plotz(4) = 0      'Stable carbon isotopes
plots(4) = 1
plotz(5) = 0       'Radioactive carbon isotopes
plots(5) = 100
'The labels are PLOTL$
plotl$(1) = "cas": plotl$(2) = "sgcs": plotl$(3) = "alks"
plotl$(4) = "dcs": plotl$(5) = "drs": plotl$(6) = "fuel"
GOSUB GRINC
RETURN
END
'****************************************************************************
PLOTTER:     'Subroutine plots values as the calculation proceeds
```

```
'The scale is normalized relative to the starting values
FOR jplot = 1 TO nrow      'Which values to plot
    ploty(jplot) = y(jplot)
NEXT jplot
ploty(6) = fuel
GOSUB PLTC
RETURN
END
'*****************************************************************************
Plus subroutines GAUSS and SWAPPER from Program DGC03
Subroutine STEPPER from Program DGC04
Subroutine SLOPER from Program DGC08
Subroutine STARTER, STOPPER, FILER, GRINC, PLTC, CARBONATE,
    and CORE from Program DGC09
Subroutine CHECKSTEP from Program ISOT01
```

DEFINITIONS in program ISOT02, are the calcium ion concentration in the waters of the lagoon, *cas,* the total dissolved carbon concentration, *sigcs,* the alkalinity, *alks,* the ^{13}C ratio, *dcs,* and the ^{14}C ratio, *drs.* Five equations describe the properties of the water of the lagoon, assumed to be homogeneous. The corresponding properties of the open ocean are held fixed at typical values for surface seawater, specified in subroutine SPECS. The partial pressure of carbon dioxide in the atmosphere is also held constant at 1 PAL. The waters of the lagoon exchange carbon with the atmosphere at the same rate, controlled by *distime,* as do the waters of the open ocean $(pco2s - pco2)/distime.$ Seawater flows into the lagoon, carrying with it the dissolved constituents of seawater at a rate equivalent to a thickness of 4 m/y, as specified at the top of subroutine SPECS. The outflow is just 2 m/y, the difference between these two numbers reflecting the influence of evaporation. The corresponding term in equation 1 is $(casw * fluxin - cas * fluxout)/depth.$ I take the depth of the water in the lagoon to be 5 meters, and so the residence time of water in the lagoon is just a few years.

Evaporation concentrates the dissolved constituents of seawater. Because the assumed inflow is twice the assumed outflow, conservative properties—properties that are not affected by precipitation, dissolution, or exchange with the atmosphere—are concentrated by a factor of 2. This increase in concentration changes the balance between the dissolved carbon species, causing the precipitation of calcium carbonate and the release of carbon dioxide into the atmosphere. My equations therefore include terms describing the exchange of carbon dioxide with the atmosphere, using the formulation previously used for the global ocean, and for the precipitation of calcium carbonate, *diss*.

I assume that the rate of precipitation of calcium carbonate depends on the degree of saturation of the waters with respect to calcium carbonate. This degree of saturation is expressed by the quantity

$$omegadel = co3s * cas/csat - 1$$

where *co3s* is the concentration of carbonate ions, *cas* is the concentration of calcium ions, and *csat* is the saturation ion product, with values specified in subroutine SPECS. The dissolution rate, which is the negative of the precipitation rate, is given by

$$diss = (\exp(-disscon * omegadel) - \exp(pcpcon * omegadel)) * disfac$$

where *disscon, pcpcon,* and *disfac* are constants specified in subroutine SPECS. The units of *diss* are moles per cubic meter per year.

The assumed values of dissolution rate are plotted in Figure 6–4. The

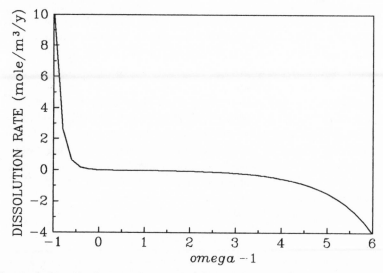

Fig. 6–4. The dissolution rate as a function of the saturation index, *omega*-1, where *omega* = *co3s* * *cas/csat* is the ion activity product divided by the saturation value.

functional form I have assumed is nonlinear and asymmetric. The dissolution rate increases more strongly than the first power of the saturation index; indeed, it increases exponentially. Precipitation increases more slowly with supersaturation than does dissolution with undersaturation. This function is intended to reflect the fact that the precipitation of calcium carbonate from supersaturated seawater can be quite slow but that the calcium carbonate exposed to undersaturated water dissolves quite quickly. I assume that no isotopic fractionation is involved in the precipitation of calcium carbonate. Precipitated calcium carbonate has the same isotopic composition as does the water of the lagoon, and the corresponding term in equation 4 is *diss* * *dcs*. The exchange of carbon dioxide with the atmosphere does, however, fractionate the carbon isotopes. Atmospheric carbon is isotopically lighter than seawater carbon, so the waters of the lagoon become isotopically heavier when carbon dioxide is driven off into the atmosphere. The term in equation 4 is

$$(pco2 * dcp - pco2s * (dcs - delcatm))/distime$$

I include one other process that might affect the properties of the lagoon's waters: biological production of isotopically light organic carbon, a process described by the term *prod* in the equations of subroutine EQUATIONS. Organic carbon fixation lowers total dissolved carbon and also increases the isotope ratio of the carbon left behind in the water. For my first experiments I set *prod* = 0 in subroutine SPECS, but later I will vary it sinusoidally to represent a seasonal change in net biological carbon fixation.

I shall now conduct a series of numerical experiments to see how this imaginary lagoon responds to seasonal changes in organic productivity, in water temperature, and in evaporation rate. These different numerical experiments are readily carried out by making minor changes in subroutine OTHER. First I make the variable *fuel* a sine function with a period of one year. Then I successively incorporate this sine function into expressions for productivity, temperature, and evaporation rate. These successive modifications have been left as comments in subroutine OTHER to make it easy to see what has been done.

6.5.1 Seasonal Change in Productivity

First I set the variable *prod* = *fuel*. This amounts to a seasonally varying biological productivity, which varies between 1 and −1 moles of carbon per cubic meter per year. I start the calculation at 1 year, with the initial values equal to the seawater values, and run it out to a time of 10 years.

The results are shown in Figures 6–5 and 6–6. Figure 6–5 depicts how the system evolves from its initial conditions to a repeatable oscillation about annual average conditions. This evolution is clearest for the calcium ion concentration, which rises toward twice the seawater value. Calcium does not quite reach twice the seawater value because it is removed from the system by the precipitation of calcium carbonate. The rise in calcium is a consequence of the evaporative concentration of the water's dissolved constituents.

The evaporative concentration increases the water's ionic strength, which affects the activity coefficients and thus the carbonate solubility and dissociation constants (Butler, 1982). I ignore this effect here, but it should be included in a more realistic simulation.

The radiocarbon ratio also evolves very rapidly from its initial value of −50 to an average value of about −8 per mil. This evolution is not a consequence of evaporative concentration but, instead, of an approach to equilibrium with atmospheric carbon dioxide. Average surface seawater contains significantly less radiocarbon than does the atmosphere because its isotopic composition is affected by exchange with the deep ocean as

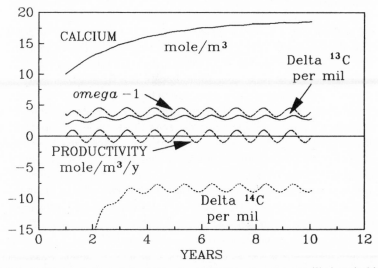

Fig. 6–5. The evolution of the lagoon's waters in response to oscillations in biological productivity. The results show the adjustment of the system from an initial composition equal to that of seawater. This figure shows isotope ratios, calcium concentration, the saturation index, and productivity.

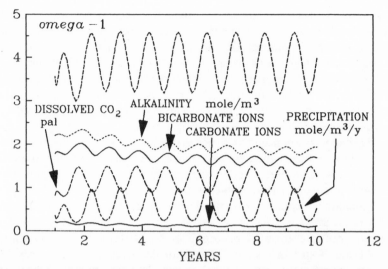

Fig. 6–6. The evolution of the lagoon's waters in response to oscillations in biological productivity. The results show the adjustment of the system from an initial composition equal to that of seawater. This figure shows dissolved carbon species, the saturation index, and the precipitation rate.

well as with the atmosphere. Radioactive decay of carbon in the deep ocean gives it a markedly lower concentration of [14]C. On the other hand, the lagoon has a depth of only 5 meters and equilibrates quite rapidly with the atmosphere. The final value for radiocarbon reflects a balance between exchange with the atmosphere at a value of 0 per mil and exchange with the surface ocean at a value of −50 per mil. The seasonal oscillation in the radiocarbon ratio is a response to the seasonal variation of productivity, plotted in Figure 6–5. Photosynthesis produces isotopically light organic carbon. When productivity is positive, therefore, the radiocarbon ratio increases. When productivity is negative, corresponding to an excess of respiration over photosynthesis, the radiocarbon ratio grows smaller.

The fluctuations in the stable isotope ratio are also plotted in Figure 6–5. These variations are parallel to the radiocarbon ratio but are of smaller amplitude because there is less fractionation for ^{13}C than for [14]C. Figure 6–5 also shows the response of the saturation state of lagoon waters to the seasonally varying productivity. Saturation state is plotted as *omega*-1, where *omega* is the ratio of the product of calcium and carbonate ion concentrations to the saturation value. The water is quite strongly supersaturated. The fluctuations in saturation value are in phase with the fluctuations in productivity. Positive productivity extracts dissolved carbon from the water, increasing the concentration of carbonate ions and therefore the saturation index. The seasonal fluctuations in calcium ion concentration are relatively small.

The saturation index, *omega*-1, is plotted again in Figure 6–6 to provide an element common to both figures. The precipitation rate (equal to minus the dissolution rate) is also plotted in this figure, its behavior being entirely a response to the variations in saturation state. The figure also shows how the different species of the carbon system respond to the seasonal fluctuation in productivity and in carbonate precipitation. Positive productivity draws down total dissolved carbon, causing more carbonate precipitation. The result is the very nearly parallel variation of alkalinity and bicarbonate ion concentrations shown in Figure 6–6. The carbonate ion concentration is out of phase and is a maximum when bicarbonate and alkalinity are close to their minimum values.

Figure 6–6 also shows the variation in the partial pressure of carbon dioxide in equilibrium with the lagoon's waters. The average value of this pressure exceeds the atmospheric value, 1, so on average, carbon dioxide is evaporating from the lagoon. The evaporation rate is greatest at times of maximum alkalinity and bicarbonate concentration and minimum carbonate ion concentration.

6.5.2 Seasonal Change in Temperature

I shall now examine the effects of seasonally varying water temperature. First, I remove the seasonal variation of productivity in subroutine OTHER and modify, instead, the value of water temperature, adding to the average value of 15°C an oscillating component equal to 5 * *fuel*. In other words, the temperature oscillates sinusoidally about an average value of 15°C with an amplitude of 5°C. This behavior is plotted in Figure 6–7. I continue the calculation from the final values achieved in the previous study, by using as starting conditions the values recorded in file ENDVAL.PRN. I have modified the expression for *xstart* at the beginning of the program to use the file value for the initial value of *x*. It is quite easy to resume a calculation by modifying *xstart* in this way and making the appropriate responses to the queries posed by subroutine FILER.

The temperature experiment can be expected to show only the effects of the temperature dependence of the equilibrium constants in the carbonate system. Other possible consequences of changing temperature are not included in the simulation. Figure 6–7 shows little response by the calcium

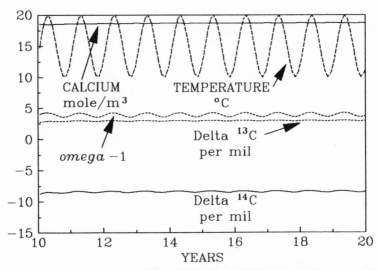

Fig. 6–7. The evolution of the lagoon's waters in response to oscillations in temperature. This figure shows isotope ratios, calcium concentration, the saturation index, and temperature.

ion concentration or carbon isotopes to the seasonal change in temperature. The saturation index does vary, being a maximum when the temperature is high. I have assumed that the solubility of calcium carbonate is independent of temperature, so this variation in saturation state and the parallel variation in precipitation rate shown in Figure 6–8 are consequences of changes in the carbonate ion concentration caused by the sensitivity to temperature of the carbonate dissociation constants.

Figure 6–8 shows how the partial pressure of carbon dioxide in equilibrium with surface water oscillates in phase with the fluctuations in precipitation rate, saturation state, and temperature. The oscillations in alkalinity and bicarbonate concentrations have shifted in phase by about 90° because these quantities decrease when precipitation and evaporation are removing carbon from the system at above-average rates.

6.5.3 Seasonal Change in Evaporation Rate

In the final experiment I simulate the effect of seasonal change in the evaporation rate by adding the sinusoidally varying *fuel* to the influx of water, *fluxin*. I leave *fluxout* unchanged at a value of 2 m/y. The influx is

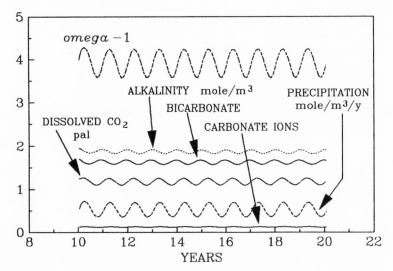

Fig. 6–8. The evolution of the lagoon's waters in response to oscillations in temperature. This figure shows dissolved carbon species, the saturation index, and the precipitation rate.

therefore varying seasonally between 5 and 3 m/y, a behavior shown in Figure 6–9. The calculation is carried on, as before, from the end of the previous calculation, with the results in Figure 6–9 depicting an oscillating response of the calcium ion concentration to the oscillation in evaporation rate. This is a straightforward effect of evaporative concentration. The radiocarbon ratio responds quite strongly, a result of the seasonal change in the amount of seawater mixed into the lagoon, a large evaporation rate corresponding to the maximum inflow of isotopically light seawater to the lagoon. By comparison, the response of the ^{13}C ratio is very small, because there is little difference between the seawater value and the value in the lagoon. Oscillations in the saturation index are 90° out of phase with the oscillations in influx. Saturation state, like calcium ion concentration, continues to increase as long as the influx and evaporation rates are above average.

Figure 6–10 shows how the various elements of the carbon system respond to the seasonal change in evaporation rate. Although the fluctuations in carbonate ion concentration cannot be seen on the scale of this figure, examination of the numbers shows that the amplitude of the carbonate fluctuation is about 3 percent, comparable to the amplitude of the

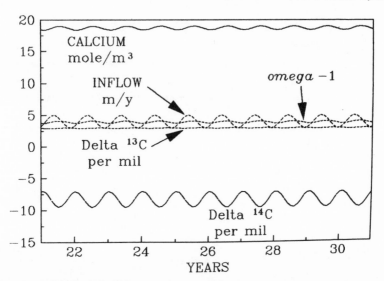

Fig. 6–9. The evolution of the lagoon's waters in response to oscillations in the evaporation rate. This figure shows isotope ratios, calcium concentration, the saturation index, and water inflow.

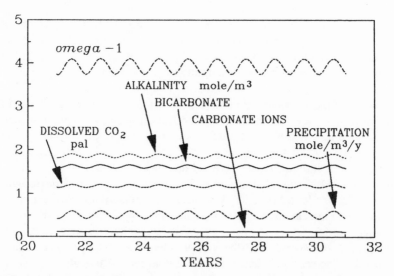

Fig. 6–10. The evolution of the lagoon's waters in response to oscillations in the evaporation rate. This figure shows dissolved carbon species, the saturation index, and the precipitation rate.

calcium ion oscillation. In this experiment, therefore, the fluctuations in saturation state are caused equally by fluctuations in calcium and carbonate concentrations, whereas in the other experiments, the calcium fluctuations were negligible.

These three numerical experiments show how the waters of an evaporating lagoon respond differently to the different seasonal perturbations that might affect them. Some record of these perturbations might, in principle, be preserved in the carbonate sediments precipitated in the lagoon. All three perturbations—productivity, temperature, and evaporation rate—cause seasonal fluctuations in the saturation state of the water and in the rate of carbonate precipitation. Temperature oscillations have little effect on the carbon isotopes. Although seasonally varying evaporation rates affect ^{14}C, they have little effect on ^{13}C. Productivity fluctuations affect both of the carbon isotopes.

6.6 Summary

In this chapter I explained how isotope ratios may be calculated from equations that are closely related, but not identical, to the equations for the bulk species. Extra terms arise in the isotope equations because isotopic composition is most conveniently expressed in terms of ratios of concentrations. I illustrated the use of these equations in a calculation of the carbon isotopic composition of atmosphere, surface ocean, and deep ocean and in the response of isotope ratios to the combustion of fossil fuels. As an alternative application, I simulated the response of the carbon system in an evaporating lagoon to seasonal changes in biological productivity, temperature, and evaporation rate. With a simulation like the one presented here it is quite easy to explore the effects of various perturbations. Although not done here, it would be easy also to examine the sensitivity of the results to such parameters as water depth and salinity.

7 Climate: A Chain of Identical Reservoirs

7.1 Introduction

One class of important problems involves diffusion in a single spatial dimension, for example, height profiles of reactive constituents in a turbulently mixing atmosphere, profiles of concentration as a function of depth in the ocean or other body of water, diffusion and diagenesis within sediments, and calculation of temperatures as a function of depth or position in a variety of media. The one-dimensional diffusion problem typically yields a chain of interacting reservoirs that exchange the species of interest only with the immediately adjacent reservoirs. In the mathematical formulation of the problem, each differential equation is coupled only to adjacent differential equations and not to more distant ones. Substantial economies of computation can therefore be achieved, making it possible to deal with a larger number of reservoirs and corresponding differential equations.

In this chapter I shall explain how to solve a one-dimensional diffusion problem efficiently, performing only the necessary calculations. The example I shall use is the calculation of the zonally averaged temperature of the surface of the Earth (that is, the temperature averaged over all longitudes as a function of latitude). I first present an energy balance climate model that calculates zonally averaged temperatures as a function of latitude in terms of the absorption of solar energy, which is a function of latitude, the emission of long-wave planetary radiation to space, which is a function of temperature, and the transport of heat from one latitude to another. This heat transport is represented as a diffusive process, dependent on the temperature gradient or the difference between temperatures in adjacent latitude bands. I use the energy balance climate model first to calculate annual average temperature as a function of latitude, comparing the calculated results with observed values and tuning the simulation by adjusting the diffusion parameter that describes the transport of energy between latitudes.

I then show that most of the elements of the *sleq* array for this problem are zero. Nonzero elements are present only on the diagonal and immediately adjacent to the diagonal. The array has this property because each differential equation for temperature in a latitude band is coupled only to temperatures in the adjacent latitude bands. The subroutine SLOPER, which calculates the elements of the *sleq* array, can be modified so that it does not waste time calculating elements that are known in advance to be zero. Similarly, the subroutine GAUSS need not take the time to convert to zero elements that are zero already. I present suitably modified versions of both these subroutines. The new solver is a lot faster than the old one.

I apply the more efficient simulation to a calculation of seasonally varying temperature as a function of latitude. In this simulation, the tuning parameter that controls the seasonal response of the temperature is the effective heat capacity of a column of atmosphere and surface. Approximately 50 meters of ocean depth is involved in the seasonal cycle of temperature, but the penetration of the seasonal temperature wave into the land surface is much less effective. The heat capacity of the land and its overlying atmosphere is therefore fifty times smaller than the heat capacity of the ocean and its overlying atmosphere. Seasonal temperature fluctuations are of larger amplitude in latitude regions with a large fraction of land than they are in latitude regions with a small fraction of land.

I use the seasonal simulation to explore the sensitivity of this energy balance climate model to such features of the climate system as permanent ice and snow at high latitudes, seasonal ice and snow, cloud cover, carbon dioxide amount, and the distribution of the continents.

7.2 Energy Balance Climate Model

The climate is an important aspect of the environment, an aspect that interacts strongly with the composition of the ocean and atmosphere. This interaction works in two ways: Climate is influenced by composition through the greenhouse effect, and climate also influences composition through its effect on reaction rates, particularly on weathering and the flux of dissolved constituents into the sea. Full-scale climate models are exceedingly complicated and can run only on supercomputers. But here I shall demonstrate how one aspect of the climate system—average temperature and its dependence on energy balance—can reasonably be studied using the computational methods developed in this book.

The simulation balances the radiant energy absorbed by a portion of the Earth's surface and the overlying atmosphere, the energy lost to space from this portion of the Earth, and the transport of energy from one portion of the Earth to another by the ocean and atmosphere. It solves for a single parameter of climate, surface temperature.

I begin by dividing the Earth's surface into eighteen latitude bands, each 10° wide, extending from the South Pole to the North Pole. These are the eighteen reservoirs of the simulation. Corresponding to these reservoirs are eighteen differential equations for the evolution of the average temperature of each latitude band. Energy fluxes in climatology are frequently expressed in units of watts per square meter of Earth surface area. I shall use these units in the simulation to be presented here. Several conversion factors result from this choice of units. First, I choose to have my calculation proceed with a time unit of years rather than seconds, so I introduce the number of seconds per year. Second, although the calculations are performed in terms of the energy content per square meter of Earth surface, the total energy content of a latitude band is this quantity multiplied by the area of the latitude band. Relative areas rather than absolute areas are what matter, and these are what I use.

Program DAV07 follows. The differential equations, as usual, appear in subroutine EQUATIONS. In each equation, the first term describes the energy exchange with the adjacent latitude bands. For the first and last reservoirs, of course, the exchange is only with the adjacent, lower-latitude reservoir, there being no higher-latitude reservoir. The second term describes the absorption of solar energy, given by incident solar flux or insolation multiplied by $1 - albedo$. *Albedo* is the fraction of the incident energy reflected off into space without being absorbed. The third term, *lwflux*, describes the emission of long-wave radiation by the atmosphere and surface to space.

The transport of heat between latitude bands is assumed to be diffusive and is proportional to the temperature difference divided by the distance between the midpoints of each latitude band. This is the temperature gradient. In this simulation all these distances are equal, so the distance need not appear explicitly. The temperature gradient is multiplied by a transport coefficient here called *diffc*, the effective diffusion coefficient. The product of the diffusion coefficient and the temperature gradient gives the energy flux between latitude zones. To find the total energy transport, we must multiply by the length of the boundary between the latitude zones. In

```
'Program DAV07 is an 18 reservoir energy balance climate model using
'the long wave radiation formulation of Kuhn et al. (1989) in LWFLUX
'Albedo formulation of Thompson and Barron in SWALBEDO
'with land and sea ice.
'Steady state annual average insolation read by CLIMINP.
'This program uses the old solver GAUSS
nrow = 18                 'the number of equations and unknowns
ncol = nrow + 1
'===================================================================
'All the dimension statements are collected here
DIM sleq(nrow, ncol), unk(nrow)
DIM y(nrow), dely(nrow), yp(nrow), incind(nrow)
'Establish arrays for graphics routines, GRAFINIT and PLOTTER
numplot = 18              'Number of variables to plot. Not more than 19
DIM ploty(numplot), plotz(numplot), plots(numplot), plotl$(numplot)
nlat = 18                         'Radiation subroutines
DIM land(nlat), oceancl(nlat), landcl(nlat), zatemp(nlat)
DIM relarea(nlat)
DIM lwflux(nlat), xlat(nlat), albedo(nlat), insol(nlat)
DIM hcap(nlat), cxlat(nlat), latmp(nlat), aveza(nlat)
'===================================================================
GOSUB SPECS
GOSUB FILER                       'Choose files for input and output
xstart = 1          'time to start
xend = 10    'time to stop
x = xstart
delx = .001                  'measure time in years
mstep = 1
GOSUB CLIMINP
GOSUB CORE
END
'*********************************************************************
SWALBEDO: 'Subroutine calculates the short wave albedo
'approximating the formulation of Thompson and Barron
FOR jlat = 1 TO nlat
'Analytic representation of dependence of albedo on daily average
'zenith angle
    IF y(jlat) < 10 THEN              'Temperature effect
        allcy = .38 + .0073 * (aveza(jlat) - 50) / (1 + (10 - y(jlat)) / 70
        allcy = allcy + (10 - y(jlat)) / 160
        IF allcy > .7 THEN allcy = .7       'upper limit on land albedo
        allcr = .17 + .0018 * (aveza(jlat) - 50)
```

```
        temporary = (1 + (y(jlat) - 10) / 20)
        allcr = allcr + .000065 * (aveza(jlat) - 50) ^ 2 * temporary
        allcr = allcr - .012 * (y(jlat) - 10)
        IF allcr > .7 THEN allcr = .7'upper limit on land albedo
    ELSE
        allcy = .377 + .0075 * (aveza(jlat) - 50)
        allcr = .17 + .0017 * (aveza(jlat) - 50)
        allcr = allcr  + .00009 * (aveza(jlat) - 50) ^ 2
    END IF
    fi = (4 - y(jlat)) / 25            'ice fraction
    IF fi > 1 THEN fi = 1
    IF fi < 0 THEN fi = 0
    alocy = .36 + .0075 * (aveza(jlat) - 50) * (1 - fi / 3.2) + fi * .145
    alocr = .102 + .0002 * (aveza(jlat) - 50)
    alocr = alocr + .00015 * (aveza(jlat) - 50) ^ 2 * (1 - fi / 1.4)
    alocr = alocr + fi * .31
    alland = allcy * landcl(jlat) + allcr * (1 - landcl(jlat))
    alocean = alocy * oceancl(jlat) + alocr * (1 - oceancl(jlat))
    albedo(jlat) = land(jlat) * alland
    albedo(jlat) = albedo(jlat) + (1 - land(jlat)) * alocean
NEXT jlat
RETURN
'*******************************************************************************
SPECS: 'Subroutine to read in the specifications of the problem
pco2 = 1
solcon = 1380              'W/m^2
diffc = 3                  'Heat transport
odhc = 35                  'Heat exchange depth in ocean in meters
hcrat = 50                 'Heat constant ratio, ocean to land
hcconst = 4185500! * odhc
secpy = 365.25 * 24 * 3600
'Temperature in degrees C.  Starting values from Sellers
y(18) = -23.4
y(17) = -15.7
y(16) = -7
y(15) = .7
y(14) = 7.7
y(13) = 14.2
y(12) = 20.6
y(11) = 25.3
y(10) = 25.7
y(9) = 25
```

```
y(8) = 23.5
y(7) = 19
y(6) = 13.7
y(5) = 8.9
y(4) = 1.4
y(3) = -10.8
y(2) = -19.3
y(1) = -47.7
'INCIND=1 for test on relative increment
FOR jlat = 1 TO nlat
    incind(jlat) = 3          'Absolute increment < 3 degrees
NEXT jlat
RETURN
'*****************************************************************************
PRINTER:    'Subroutine writes a file for subsequent plotting
FOR jrow = 1 TO nlat
    PRINT #1, y(jrow);
NEXT jrow
PRINT #1, x
GOSUB PLOTTER
RETURN
'*****************************************************************************
EQUATIONS: 'These equations define the problem
tvar = x + delx
GOSUB OTHER
yp(1) = diffc * cxlat(2) / relarea(1) * (y(2) - y(1))
yp(1) = (yp(1) + insol(1) * (1 - albedo(1)) - lwflux(1)) * secpy / hcap(1)
FOR jlat = 2 TO nlat - 1
yp(jlat) = cxlat(jlat + 1) * (y(jlat + 1) - y(jlat))
yp(jlat) = yp(jlat)  - cxlat(jlat) * (y(jlat) - y(jlat - 1))
yp(jlat) = yp(jlat) * diffc / relarea(jlat)
yp(jlat) = yp(jlat) + insol(jlat) * (1 - albedo(jlat)) - lwflux(jlat)
yp(jlat) = yp(jlat) * secpy / hcap(jlat)
NEXT jlat
yp(nlat) = -diffc * cxlat(nlat) / relarea(nlat) * (y(nlat) - y(nlat - 1))
yp(nlat) = yp(nlat) + insol(nlat) * (1 - albedo(nlat)) - lwflux(nlat)
yp(nlat) = yp(nlat) * secpy / hcap(nlat)
RETURN
'*****************************************************************************
OTHER: 'Subroutine evaluates other quantities that change with time
GOSUB DEFINITIONS
GOSUB LWRAD
```

```
        GOSUB SWALBEDO
        RETURN
'******************************************************************************
LWRAD:   'Subroutine calculates the longwave flux as a function of
         'temperature and CO2 pressure.  Formulation of Kuhn et al.
        alp = LOG(pco2 * 280 / 345)
        aclear = 180.25 - 2.9453 * alp - .32122 * alp * alp
        acloud = 124.29 - 1.5495 * alp - .05835 * alp * alp
        bclear = 2.2301 - .019822 * alp - .010285 * alp * alp
        bcloud = 1.8486 - .0235 * alp - .0027448 * alp * alp
        cclear = .004863 + .0003722 * alp - .000093138# * alp * alp
        ccloud = .0081951 + .000033857# * alp - .000048093# * alp * alp
        FOR jlat = 1 TO nlat
            foc = 1! - land(jlat)
            fcloud = foc * oceancl(jlat) + land(jlat) * landcl(jlat)
            arad = fcloud * acloud + (1! - fcloud) * aclear
            brad = ·fcloud * bcloud + (1! - fcloud) * bclear
            crad = fcloud * ccloud + (1! - fcloud) * cclear
            lwflux(jlat) = arad + brad * (zatemp(jlat) + 23)
            lwflux(jlat) = lwflux(jlat) + crad * (zatemp(jlat) + 23) ^ 2
        NEXT jlat
        RETURN
'******************************************************************************
DEFINITIONS:
        FOR jlat = 1 TO nlat
            zatemp(jlat) = y(jlat)
        NEXT jlat
        RETURN
'******************************************************************************
CLIMINP:         'Subroutine reads land area, sin(latitude), and other
         'parameters of the climate system
        OPEN "CLIMINP.PRN" FOR INPUT AS #5
        FOR jvar = 1 TO nlat
            INPUT #5, latmp(jvar), xlat(jvar), relarea(jvar), land(jvar)
            INPUT #5, landcl(jvar), oceancl(jvar), insol(jvar), aveza(jvar)
        NEXT jvar
        CLOSE #5
        FOR jlat = 1 TO nlat
            insol(jlat) = insol(jlat) * solcon / 1400    'tune solar luminosity
            hcap(jlat) = hcconst * ((1 - land(jlat)) + land(jlat) / hcrat)
            'Assume heat capacity of land is 1/hcrat times that of ocean
            cxlat(jlat) = SQR(1 - xlat(jlat) ^ 2)
```

```basic
    latmp(jlat) = latmp(jlat) / 180 * pi          'latitude in radians
NEXT jlat
RETURN
'***************************************************************************
GRAFINIT:                  'Initialize graphics
'Plot relative departures from PLOTZ
'Sensitivity of the plot depends on PLOTS
'The labels are PLOTL$
FOR jlat = 1 TO numplot
    plotz(jlat) = y(jlat)
    plots(jlat) = 4
    plotl$(jlat) = STR$(jlat)
NEXT jlat
GOSUB GRINC
RETURN
END
'***************************************************************************
PLOTTER:        'Subroutine plots values as the calculation proceeds
'The scale is normalized relative to the starting values
FOR jplot = 1 TO nrow          'Which values to plot
    ploty(jplot) = y(jplot)
NEXT jplot
GOSUB PLTC
RETURN
END
'***************************************************************************
CORE:    'Subroutine directs the calculation
maxinc = .1
dlny = .001
count = 0                   'count time steps
GOSUB GRAFINIT
GOSUB PRINTER
DO WHILE x < xend
    count = count + 1
    GOSUB SLOPER
    GOSUB GAUSS
    FOR jrow = 1 TO nrow
        dely(jrow) = unk(jrow)
    NEXT jrow
    GOSUB CHECKSTEP
LOOP
PRINT #1,
```

```
FOR jrow = 1 TO nrow
    PRINT #1, aveza(jrow); lwflux(jrow); albedo(jrow); insol(jrow)
NEXT jrow
GOSUB STOPPER
RETURN
'************************************************************************
Plus subroutines GAUSS and SWAPPER from Program DGC03
Subroutine STEPPER from Program DGC04
Subroutine SLOPER from Program DGC08
Subroutine STARTER, STOPPER, FILER, GRINC, PLTC from Program DGC09
Subroutine CHECKSTEP from Program ISOT01
```

subroutine EQUATIONS this length is expressed in relative units by *cxlat*, the cosine of the latitude of the bounding line. Finally, to convert to heat transport per unit area of reservoir, it is necessary to divide by *relarea*— the area of the reservoir in relative units—the difference between the sines of the latitudes of the two bounding lines.

In subroutine EQUATIONS, therefore, *yp* is the rate of change of temperature, *y* is the temperature, *diffc* is the diffusion coefficient, *cxlat* is the relative length of the boundary line between reservoirs, and *relarea* is the relative area of the reservoir. To save computational effort, the fixed functions of latitude like *cxlat* and *relarea* are read from a file by subroutine CLIMINP.

The solar energy absorbed is equal to insolation times 1 − *albedo*. In this first calculation, insolation is a function only of latitude. It is contained in the array *insol* and is read from a file by subroutine CLIMINP. Annual average insolation is plotted in Figure 7–1.

The albedo depends on surface properties—whether ocean, land, or ice—on the presence or absence of clouds, and on the zenith angle of the sun. The formulation I use is based on a detailed study by Thompson and Barron (1981). I have fitted to the results of their theory the analytical expressions contained in subroutine SWALBEDO. Figures 7–2 and 7–3 illustrate the calculated albedos for various conditions: Figure 7–2 shows the variation of albedo for clear and cloudy skies over land and ocean as a function of the daily average solar zenith angle, results that were calculated using subroutine SWALBEDO. The temperature was taken to be warm enough to eliminate ice and snow. The most important parameter is cloud cover, because the difference between land and ocean is most marked

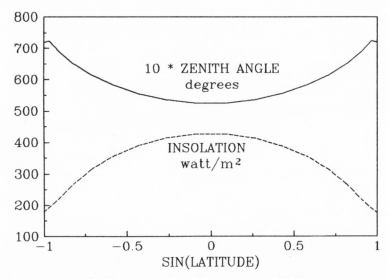

Fig. 7–1. The annual average insolation and average zenith angle as a function of the sine of latitude. The zenith angle has been multiplied by a factor of 10 so that its variation can be seen. The sine of latitude is used as the ordinate in all of these plots because it reflects the relative surface area at each latitude.

under cloudless conditions, when sunlight can directly penetrate the ocean, where it may be absorbed. Albedo increases with increasing solar zenith angle, as the sun's rays enter more nearly parallel to the Earth's surface.

In this simulation I follow Kuhn, Walker, and Marshall (1989) in taking the fraction of the sky covered with clouds to be different for ocean and for land but, in each case, a fixed function of latitude. These functions are specified by the arrays *oceancl* and *landcl* read from a file by subroutine CLIMINP, and they are listed in Table 7-1. The fraction of each latitude band that is land is specified by array *land,* also read by CLIMINP. The daily average solar zenith angle is a function of latitude and season. For this annually averaged calculation I have taken the daily average zenith angle, weighted it by the daily average insolation, and averaged it over an entire year to calculate the annual average of the daily average solar zenith angle This zenith angle, plotted in Figure 7–1, varies from 50° on the equator to just over 70° at the poles. It is as large as 50° at the equator because the quantity is averaged throughout the daylight hours and refers not just to the middle of the day when the zenith angle is nearly zero. Moreover, because the zenith angle is used to describe the incidence of

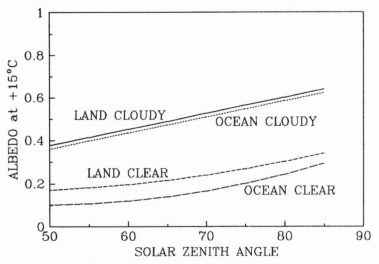

Fig. 7–2. The albedo as a function of the daily average solar zenith angle comparing clear and cloudy land and ocean. The temperature for these calculations was taken as +15°C to suppress ice and snow. In this formulation, the albedo is independent of temperature at temperatures above 10°C.

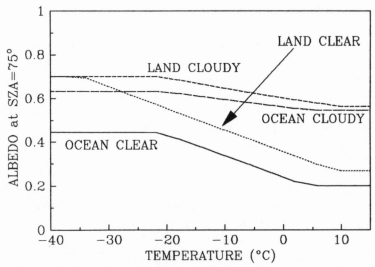

Fig. 7–3. The albedo as a function of temperature at a solar zenith angle of 75°, comparing clear and cloudy ocean and land. The temperature effect results from the large albedo of ice and snow. The sensitivity is small for cloud-covered land and ocean, larger for clear ocean, and largest for clear land.

Table 7–1. Climatic Input Parameters

Latitude	Relative area	Land fraction	Land clouds	Ocean clouds	Insolation	Zenith angle
−85	0.015192	1	0.43	0.66	170.1	71.2
−75	0.045115	0.754	0.54	0.76	187.5	71.6
−65	0.073667	0.104	0.64	0.82	211.8	68.7
−55	0.09998	0.008	0.64	0.82	252.9	64.6
−45	0.123256	0.03	0.53	0.74	301.5	61.1
−35	0.142787	0.112	0.46	0.64	342.0	58.0
−25	0.157979	0.231	0.36	0.56	373.8	55.4
−15	0.168371	0.22	0.49	0.54	397.0	53.6
− 5	0.173648	0.236	0.65	0.58	409.1	52.6
5	0.173648	0.228	0.65	0.58	409.1	52.6
15	0.168371	0.264	0.49	0.54	397.0	53.6
25	0.157979	0.376	0.36	0.56	373.8	55.5
35	0.142787	0.428	0.46	0.64	342.0	58.1
45	0.123256	0.525	0.53	0.74	301.5	61.2
55	0.09998	0.572	0.64	0.82	252.9	64.7
65	0.073667	0.706	0.64	0.82	211.8	68.8
75	0.045115	0.287	0.54	0.76	187.5	71.7
85	0.015192	0.066	0.43	0.66	170.1	71.2

sunlight on a horizontal surface, the average is actually performed on the cosine of the zenith angle. On the other hand, the average zenith angle at the poles is not larger than 72° because it is weighted by insolation. Values greater than 90° during the polar night receive no weight. Large solar zenith angles correspond to small insolation and so receive little weight. In this annual average simulation, the annual average solar zenith angle is represented by array *aveza,* read by CLIMINP.

The parameters and results of the albedo calculation are illustrated in Figure 7–4, which shows the calculated albedo for annual average temperatures, which we shall describe later. Figure 7–4 also indicates the fixed values of land fraction, cloud cover on land, and cloud cover over the ocean that enter into the albedo calculation. Quantities are plotted against the sine of latitude because this quantity is proportional to the area of the globe at each latitude.

There is a potentially important influence of ice and snow at the surface, for ice and snow have high albedos. Ice–albedo feedback may increase the sensitivity of the climate system. That is, low temperature causes more ice and a higher albedo, allowing less absorption of sunlight and therefore causing a still lower temperature. This temperature effect is included by

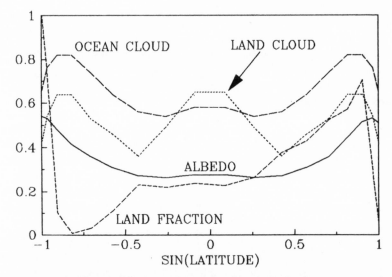

Fig. 7–4. Specified distributions of land fraction, land cloud cover, and ocean cloud cover as functions of the sine of latitude. Also plotted is the albedo calculated for these values of land and cloud and for the temperatures of Figure 7–6. Negative values of SIN(LATITUDE) correspond to the Southern Hemisphere.

functions of the temperature in subroutine SWALBEDO. Expressions are based on the derivation of Thompson and Barron (1981) but are analytical fits to the results of their theory. On land, if the temperature is lower than 10°C, albedo will increase with the decreasing temperature to a maximum value of 0.7. Ocean albedo depends on the fraction of the zone occupied by ice, a fraction expressed by a simple linear function of temperature in subroutine SWALBEDO. There will be no sea ice in this simulation if the temperature exceeds 4°C. The ocean will be completely covered with ice if the temperature is lower than −29°C. The influence of temperature on albedo at a large solar zenith angle is illustrated in Figure 7–3. The ice effect is not likely to play a role at small zenith angles because sufficiently low temperatures are not generally encountered when the average solar zenith angle is small. Subroutine SWALBEDO calculates the zonally averaged albedo as a function of latitude in terms of the temperature, land fraction, and cloud distributions, all expressed as functions of latitude. The albedo is returned in array *albedo*. Subroutine SWALBEDO is called from subroutine OTHER, which in turn is called from subroutine EQUATIONS.

The formulation of outgoing long-wave radiation is based on the results of Kuhn, Walker, and Marshall (1989). Outgoing radiation depends on cloud cover because cold, high-altitude clouds radiate less effectively than does the warm ground. Outgoing radiation depends on the temperature of the ground and also on the greenhouse effect, represented here by the partial pressure of carbon dioxide. Large amounts of carbon dioxide cause the outgoing radiation to escape to space from higher and colder levels of the atmosphere. Therefore the outgoing radiation at a given ground temperature is reduced for a larger amount of carbon dioxide, with the results illustrated in Figure 7–5. The equations in subroutine LWRAD are an analytical fit to the results of a full one-dimensional radiative convective calculation of outgoing long-wave flux as a function of temperature, cloud cover, and carbon dioxide amount. This calculation assumed a constant relative humidity, and so increasing temperature leads to increasing water vapor in the atmosphere and a larger water vapor greenhouse effect. Subroutine LWRAD calculates outgoing flux *lwflux* as a function of ground temperature, contained in the array *zatemp*, atmospheric carbon dioxide *pco2*, specified in subroutine SPECS, and land fraction, ocean cloud cover, and land cloud cover as functions of latitude read by subroutine CLIMINP.

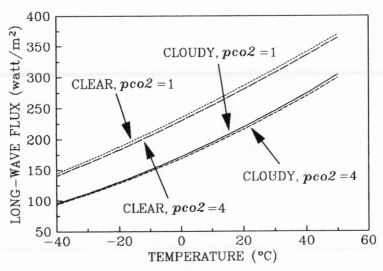

Fig. 7–5. Outgoing long-wave planetary radiation as a function of temperature, comparing clear and cloudy skies for carbon dioxide partial pressure equal to one and four times the current level.

Other parameters of the simulation are specified in subroutine SPECS. The quantity *solcon* is the solar constant, available here for tuning within observational limits of uncertainty. The quantity *diffc* is the heat transport coefficient, a freely tunable parameter. The quantity *odhc* is the depth in the ocean to which the seasonal temperature variation penetrates. In this annual average simulation, it simply controls how rapidly the temperature relaxes into a steady-state value. In the seasonal calculations carried out later in this chapter it controls the amplitude of the seasonal oscillation of temperature. The quantity *hcrat* is the amount by which ocean heat capacity is divided to get the much smaller effective heat capacity of the land. The quantity *hcconst* converts the heat exchange depth of the ocean into the appropriate units for calculations in terms of watts per square meter. The quantity *secpy* is the number of seconds in a year.

The calculation is performed in terms of degrees Celsius, including values both above and below zero. It is not convenient, therefore, to use the relative increment of temperature as a test for step size in subroutine CHECKSTEP. I use absolute increments instead. At the end of subroutine SPECS, I set *incind* equal to 3 for all equations, limiting the absolute increment in temperature to 3° per time step. Zonally averaged heat capacity as a function of latitude is calculated in subroutine CLIMINP in terms of land fraction and the heat capacity parameters specified in SPECS. It is returned in the array *hcap*.

For the real-time graphics I have chosen to plot departures of temperature from the initial values with a sensitivity of 1°, specifications that are contained in subroutines GRAFINIT and PLOTTER. I have modified subroutine CORE so that the last thing the program does before calling subroutine STOPPER is print to the RESULTS file the parameters of the energy balance, average solar zenith angle, long-wave flux, albedo, and insolation. This listing can be used to plot components of the energy balance as it exists at the end of the calculation.

7.3 Annual Average Temperature as a Function of Latitude

I start program DAV07 at 1 year of simulation time and run it to 10 years of simulation time. A time step of less than a day is necessary initially, as the system adjusts rapidly from its starting values. The time step increases as the system approaches steady state. I tune the simulation by adjusting the value of *diffc*, the transport parameter, and *solcon*, the solar constant, to

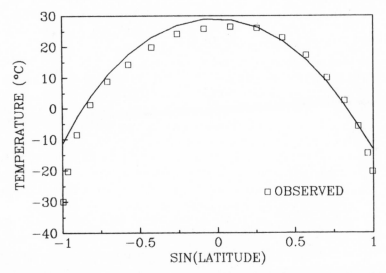

Fig. 7–6. Calculated values of the annual average temperature as a function of the sine of latitude, plotted as a solid line, compared with observed values, plotted as squares.

achieve the values illustrated in Figure 7–6. The squares in this figure are annual averaged values of zonal average temperatures from Fairbridge (1967). The solid line is the calculated temperature profile, and the required value of solar constant is within observational limits. An alternative tuning parameter would be some aspect of the system controlling albedo.

The agreement between theory and observation shown in Figure 7–6 is reasonable, considering how simple the theory is. In particular, energy transport in the real climate system is probably a more complicated function of temperature than I have assumed. The diffusive energy transport parameter is taken here to be independent of latitude, temperature, or season. Other parameterizations might improve the fit between theory and observation but might still be hard to justify on physical grounds. In the following, it is important to keep in mind that I shall be experimenting with the response of a particular simulation of the climate, and so the results may not apply to the real climate system.

The temperature values plotted in Figure 7–6 were used to calculate the albedo values plotted in Figure 7–4, but temperature influences albedo in this formulation only in its effect on low-temperature ice and snow. The components of the energy balance corresponding to the temperatures of

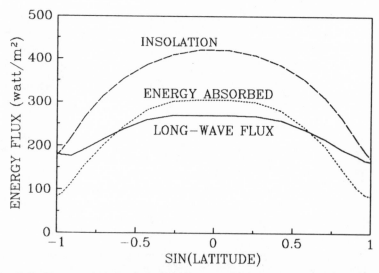

Fig. 7–7. The energy balance for the results plotted in Figure 7–6. This figure shows insolation as a function of the sine of latitude, energy absorbed by the Earth–atmosphere system, and outgoing long-wave flux. Transport in atmosphere and ocean carries excess energy from low latitudes to high latitudes. The outgoing long-wave flux is less than the energy absorbed at low latitudes and greater than that at high latitudes.

Figure 7–6 are plotted in Figure 7–7. The absorbed energy falls off more rapidly with latitude than does the long-wave flux. Transport in this climate system carries excess energy away from the tropics to higher latitudes where there is a deficit in the energy budget.

7.4 How to Avoid Unnecessary Computation

In this system, the rate of temperature change at a given latitude depends on the temperature at that latitude, which influences albedo and thus energy absorption as well as outgoing long-wave radiation, and on the temperatures in the adjacent latitude bands, which control the diffusive transport of heat in and out of the latitude band in question. There is no direct coupling to more distant reservoirs, as may be confirmed by inspection of subroutine EQUATIONS in program DAV07. The array *sleq* describes how sensitive a rate of change is to a small adjustment in the values of the dependent variables, y. Therefore, the *sleq* array has nonzero elements only on the diagonal and immediately adjacent to the diagonal. All other elements are zero. The *sleq* array looks like the following in which X represents nonzero values:

```
X X 0 0 0 0 0 . . . X
X X X 0 0 0 0 . . . X
0 X X X 0 0 0 . . . X
0 0 X X X 0 0 . . . X
0 0 0 X X X 0 . . . X
```

Obviously, then, I can save time and work by modifying subroutine SLOPER to calculate only the nonzero elements of the array. The modified subroutine appears in program DAV08 and is called subroutine SLOPERD, with the D representing diagonal. The actual calculations in SLOPERD are identical to those in SLOPER, but the limits on the FOR loops have been adjusted so that calculations are performed only for the elements on the diagonal and immediately adjacent to the diagonal.

Even more savings are possible in subroutine GAUSS, which solves the system of simultaneous linear algebraic equations described by the *sleq* array. The modified version of GAUSS appears in program DAV08 as subroutine GAUSSD. The solution of the system of linear equations begins by multiplying each equation by constant values and adding the equations together in order to achieve a matrix with zero elements below the diagonal and ones on the diagonal. The unknown solution of the system of linear algebraic equations can then be readily calculated by means of back substitution, a solution discussed originally in Chapter 3. The modifications here take advantage of the many zeros that already exist below the diagonal as well as the zeros above it. The first step is to divide the top row of the matrix, which corresponds to the first differential equation, by its first nonzero element, which in this case is its first element. Because this row has only two nonzero elements in it, this division is equivalent to dividing the second element by the first element and setting the first element to one. No further change in the first row is needed.

The next stage of the solution consists of subtracting a multiple of the first row from the lower rows, to make the first element in each of these lower rows equal to zero. But the only nonzero elements in the first column are in the first two rows, so this manipulation need extend no further than the second row. The values in the first row are multiplied by the first element in the second row, and the resultant values are subtracted from the second row to convert the first element in the second row to zero. But

```
'Program DAV08 is an 18 reservoir energy balance climate model
'using the long wave radiation formulation of Kuhn et al. (1989) in LWFLUX
'Albedo formulation of Thompson and Barron in SWALBEDO
'with land and sea ice.
'Steady state annual average insolation read by CLIMINP.
'This program uses the new solver GAUSSD and SLOPERD
nrow = 18          'the number of equations and unknowns
ncol = nrow + 1
'=======================================================================
'All the dimension statements are collected here
DIM sleq(nrow, ncol), unk(nrow)
DIM y(nrow), dely(nrow), yp(nrow), incind(nrow)
'Establish arrays for graphics routines, GRAFINIT and PLOTTER
numplot = 18       'Number of variables to plot.  Not more than 19
DIM ploty(numplot), plotz(numplot), plots(numplot), plotl$(numplot)
nlat = 18          'Radiation subroutines
DIM land(nlat), oceancl(nlat), landcl(nlat), zatemp(nlat)
DIM relarea(nlat)
DIM lwflux(nlat), xlat(nlat), albedo(nlat), insol(nlat)
DIM hcap(nlat), cxlat(nlat), latmp(nlat), aveza(nlat)
'=======================================================================
GOSUB SPECS
GOSUB FILER              'Choose files for input and output
xstart = 1       'time to start
xend = 10  'time to stop
x = xstart
delx = .001           'measure time in years
mstep = 1
GOSUB CLIMINP
GOSUB CORE
END
'***********************************************************************
CORE:  'Subroutine directs the calculation
maxinc = .1
dlny = .001
count = 0          'count time steps
GOSUB GRAFINIT
GOSUB PRINTER
DO WHILE x < xend
    count = count + 1
    GOSUB SLOPERD
    GOSUB GAUSSD
```

sonal Cycle

te, annual average temperature profile is not a very exciting
f a model that is intrinsically able to calculate changes with
ection I apply the simulation to a calculation of the seasonal
mperature as a function of latitude. The program is listed as

mportant addition to the program is the subroutine SEASON,
ates the seasonal variation of daily average insolation as a
latitude. The expressions for this calculation are given by
) and by Berger (1978). From the day of the year, this sub-
lates the true anomaly or angular distance of the Earth from
n of its orbit, the ratio of Earth–Sun distance to the semimajor
bit, and the solar declination, or latitude of the subsolar point.
quantities, the subroutine then calculates the flux of solar
ent on a horizontal area of ground and atmosphere, averaged
, and the cosine of the solar zenith angle, also averaged over
daily average zenith angle is used in subroutine SWALBEDO
albedo. The trigonometric expressions used in these calculate straightforward, although some care is needed in handling
gions where there are 24-hour days in summer and 24-hour
nter. Figure 7–8 shows insolation as a function of time of year
latitudes in the Southern Hemisphere. Northern Hemisphere
imilar but are shifted six months in phase and are of slightly
gnitude because of the eccentricity of the Earth's orbit, which
Earth closest to the sun in January.

he SEASON is called from subroutine OTHER. The time vari-
defined in subroutine EQUATIONS for this calculation and in
PRINTER to print the results. I use an additional file for listing
which is opened at the end of subroutine CLIMINP. I put it there
like CLIMINP, is peculiar to the climate problem. Subroutine
writes parameters of the climate system to this extra file. In this
ist the albedo values in RESSEAS.PRN, but for some applica-
ht list insolation or long-wave flux. The reason for this extra file
port my RESULTS file into a spreadsheet program for study,
on, and plotting. The import routine cannot handle lines longer
SCII characters, and the list of eighteen temperature values uses
f this allowance. The easiest way to list additional parameters
ternating lines, which makes plotting difficult, is to write a

```
        FOR jrow = 1 TO nrow
            dely(jrow) = unk(jrow)
        NEXT jrow
        GOSUB CHECKSTEP
LOOP
PRINT #1,
FOR jrow = 1 TO nrow
    PRINT #1, aveza(jrow); lwflux(jrow); albedo(jrow); insol(jrow)
NEXT jrow
GOSUB STOPPER
RETURN
'****************************************************************************
SLOPERD: REM Subroutine to calculate the coefficient matrix, SLEQ
'for a tri-diagonal matrix
GOSUB EQUATIONS                    'calculate the derivatives
FOR jrow = 1 TO nrow
    sleq(jrow, ncol) = yp(jrow)
NEXT jrow
FOR jcol = 1 TO nrow
    IF y(jcol) = 0 THEN yinc = dlny ELSE yinc = y(jcol) * dlny
    y(jcol) = y(jcol) + yinc
    GOSUB EQUATIONS
    jcll = jcol - 1: jcul = jcol + 1'Limits on JROW
    IF jcll < 1 THEN jcll = 1
    IF jcul > nrow THEN Jcul = nrow
    FOR jrow = jcll TO jcul     'differentiate with respect to Y(JCOL)
            sleq(jrow, jcol) = -(yp(jrow) - sleq(jrow, ncol)) / yinc
    NEXT jrow
    y(jcol) = y(jcol) - yinc   'restore the original value
    sleq(jcol, jcol) = sleq(jcol, jcol) + 1 / delx
            'extra term in diagonal elements
NEXT jcol
RETURN
REM ****************************************************************************
GAUSSD: 'Subroutine GAUSSD solves a system of simultaneous linear
'algebraic equations by Gaussian elimination and back substitution.
'The number of equations (equal to the number of unknowns) is NROW.
'The coefficients are in array SLEQ(NROW,NROW+1), where the last column
'contains the constants on the right hand sides of the equations.
'The answers are returned in the array UNK(NROW).
'This subroutine assumes that the matrix is tri-diagonal
diag = sleq(1, 1)
```

```
sleq(1, 2) = sleq(1, 2) / diag
sleq(1, ncol) = sleq(1, ncol) / diag
sleq(1, 1) = 1
coeff1 = sleq(2, 1)
sleq(2, 2) = sleq(2, 2) - sleq(1, 2) * coeff1
sleq(2, ncol) = sleq(2, ncol) - sleq(1, ncol) * coeff1
sleq(2, 1) = 0
FOR jrow = 2 TO nrow - 1
    jr = jrow + 1
    diag = sleq(jrow, jrow)
    REM divide by coefficient on the diagonal
    sleq(jrow, jr) = sleq(jrow, jr) / diag
    sleq(jrow, ncol) = sleq(jrow, ncol) / diag
    sleq(jrow, jrow) = 1
    coeff1 = sleq(jr, jrow)
    'zeroes below the diagonal
    sleq(jr, jr) = sleq(jr, jr) - sleq(jrow, jr) * coeff1
    sleq(jr, ncol) = sleq(jr, ncol) - sleq(jrow, ncol) * coeff1
    sleq(jr, jrow) = 0
NEXT jrow
sleq(nrow, ncol) = sleq(nrow, ncol) / sleq(nrow, nrow)
sleq(nrow, nrow) = 1
REM calculate unknowns by back substitution
unk(nrow) = sleq(nrow, ncol)
FOR jrow = nrow - 1 TO 1 STEP -1
    jr = jrow + 1
    unk(jrow) = sleq(jrow, ncol) - unk(jr) * sleq(jrow, jr)
NEXT jrow
RETURN
'*********************************************************************************
Plus subroutine STEPPER from Program DGC04
Subroutine STARTER, STOPPER, FILER, GRINC, PLTC from Program DGC09
Subroutine CHECKSTEP from Program ISOT01
Subroutine SWALBEDO, SPECS, PRINTER, EQUATIONS, OTHER, LWRAD, DEFINITIONS
    CLIMINP, GRAFINIT, PLOTTER from Program DAV07
```

because nonzero elements in the
and last columns, this manipulati
last columns. The first element i
element in the second row is not a
the first row.

The next step is to divide the nc
first nonzero element in the second
columns. The second element in th
tion proceeds in this way from row
subtractions required to convert th
below the diagonal and ones on t
finally been achieved, there are on
only in the last column and for the
diagonal.

$$
\begin{array}{ccccc}
1 & X & 0 & 0 & 0 \\
0 & 1 & X & 0 & 0 \\
0 & 0 & 1 & X & 0 \\
0 & 0 & 0 & 1 & X \\
0 & 0 & 0 & 0 & 1 \\
\end{array}
$$

In the procedure of back substitution
of the unknowns are calculated, it is r
unknown. Earlier unknowns do not er
this stage of the calculation is a lot s

Subroutine SWAPPER was introduc
zero element on the diagonal, rearrang
rid of a zero element on the diagonal
that zero element. A one-dimensional
element on the diagonal because the
dependent variable in the differential e
thermore, if we were to change the
would no longer have its diagonal form
has been eliminated from program DA'

Program DAV08 yields results that
results of program DAV07. There are
The new program is performing exactly
tities not equal to zero; the calculation is
program with GAUSSD and SLOPERD

7.5 The Sea

The steady-st
application o
time. In this
variation of t
DAV09.

The most i
which calcul
function of
Sellers (1965
routine calcu
the perihelio
axis of the o
From these
energy incid
over the day
the day. The
to calculate
tions are qu
the polar re
nights in wi
for several
results are
different ma
places the
Subrouti
able *tvar* is
subroutine
the results
because it,
PRINTER
example I
tions I mig
is that I in
manipulati
than 240 A
up most o
without a

```
'Program DAV09 is an 18 reservoir energy balance climate model
'using the long wave radiation formulation of Kuhn et al. (1989) in LWFLUX
'Albedo formulation of Thompson and Barron in SWALBEDO
'with land and sea ice.
'Seasonal variation of insolation and zenith angle from subroutine SEASON
'This program uses the new solver GAUSSD and SLOPERD
nrow = 18        'the number of equations and unknowns
ncol = nrow + 1
'====================================================================
'All the dimension statements are collected here
DIM sleq(nrow, ncol), unk(nrow), excoeff(nrow, ncol), ovol(nrow)
DIM y(nrow), dely(nrow), yp(nrow), incind(nrow)
'Establish arrays for graphics routines, GRAFINIT and PLOTTER
numplot = 18        'Number of variables to plot. Not more than 19
DIM ploty(numplot), plotz(numplot), plots(numplot), plotl$(numplot)
nlat = 18           'Radiation subroutines
DIM land(nlat), oceancl(nlat), landcl(nlat), zatemp(nlat)
DIM relarea(nlat)
DIM lwflux(nlat), xlat(nlat), albedo(nlat), insol(nlat)
DIM hcap(nlat), cxlat(nlat), latmp(nlat), aveza(nlat)
'Constants of the seasonal variation of insolation
CONST pi = 3.141593, obliq = 23.45 / 180 * pi, ecc = .017
CONST dayp = 3      'Day number of perihelion
'====================================================================
GOSUB SPECS
GOSUB FILER             'Choose files for input and output
xstart = 1      'time to start
xend = 4    'time to stop
x = xstart
delx = .001         'measure time in years
mstep = .1
GOSUB CLIMINP
GOSUB CORE
END
'********************************************************************
SPECS: 'Subroutine to read in the specifications of the problem
pco2 = 1
solcon = 1380       'W/m^2
diffc = 3           'Heat transport
odhc = 35           'Heat exchange depth in ocean in meters
hcrat = 50          'Heat constant ratio, ocean to land
hcconst = 4185500! * odhc
```

```
secpy = 365.25 * 24 * 3600
'Temperature in degrees C.  Starting values from Sellers
y(18) = -23.4
y(17) = -15.7
y(16) = -7
y(15) = .7
y(14) = 7.7
y(13) = 14.2
y(12) = 20.6
y(11) = 25.3
y(10) = 25.7
y(9) = 25
y(8) = 23.5
y(7) = 19
y(6) = 13.7
y(5) = 8.9
y(4) = 1.4 ·
y(3) = -10.8
y(2) = -19.3
y(1) = -47.7
'INCIND=1 for test on relative increment
FOR jlat = 1 TO nlat
    incind(jlat) = 3     'Absolute increment < 3 degrees
NEXT jlat
RETURN
'***************************************************************************
PRINTER:  'Subroutine writes a file for subsequent plotting
tvar = x                'time for seasonal change
GOSUB OTHER              'print current values
FOR jrow = 1 TO nlat
    PRINT #1, y(jrow);
    PRINT #5, albedo(jrow);
NEXT jrow
PRINT #1, x
PRINT #5, x
GOSUB PLOTTER
RETURN
'***************************************************************************
EQUATIONS: 'These equations define the problem
tvar = x + delx         'Time for seasonal change
GOSUB OTHER
yp(1) = diffc * cxlat(2) / relarea(1) * (y(2) - y(1))
```

```
yp(1) = (yp(1) + insol(1) * (1 - albedo(1)) - lwflux(1)) * secpy / hcap(1)
FOR jlat = 2 TO nlat - 1
yp(jlat) = cxlat(jlat + 1) * (y(jlat + 1) - y(jlat))
yp(jlat) = yp(jlat)  - cxlat(jlat) * (y(jlat) - y(jlat - 1))
yp(jlat) = yp(jlat) * diffc / relarea(jlat)
yp(jlat) = yp(jlat) + insol(jlat) * (1 - albedo(jlat)) - lwflux(jlat)
yp(jlat) = yp(jlat) * secpy / hcap(jlat)
NEXT jlat
yp(nlat) = -diffc * cxlat(nlat) / relarea(nlat) * (y(nlat) - y(nlat - 1))
yp(nlat) = yp(nlat) + insol(nlat) * (1 - albedo(nlat)) - lwflux(nlat)
yp(nlat) = yp(nlat) * secpy / hcap(nlat)
RETURN
'*************************************************************************
OTHER: 'Subroutine evaluates other quantities that change with time
GOSUB DEFINITIONS
GOSUB LWRAD
GOSUB SEASON
GOSUB SWALBEDO
RETURN
'*************************************************************************
CLIMINP:    'Subroutine reads land area, sin(latitude), and other
'parameters of the climate system
OPEN "CLIMINP.PRN" FOR INPUT AS #5
FOR jvar = 1 TO nlat
    INPUT #5, latmp(jvar), xlat(jvar), relarea(jvar), land(jvar)
    INPUT #5, landcl(jvar), oceancl(jvar), insol(jvar), aveza(jvar)
NEXT jvar
CLOSE #5
FOR jlat = 1 TO nlat
    insol(jlat) = insol(jlat) * solcon / 1400     'tune solar luminosity
    hcap(jlat) = hcconst * ((1 - land(jlat)) + land(jlat) / hcrat)
    'Assume heat capacity of land is 1/hcrat times that of ocean
    cxlat(jlat) = SQR(1 - xlat(jlat) ^ 2)
    latmp(jlat) = latmp(jlat) / 180 * pi        'latitude in radians
NEXT jlat
'Output seasonally varying parameters
OPEN "RESSEAS.PRN" FOR OUTPUT AS #5
RETURN
'*************************************************************************
SEASON: 'Program SEASON calculates seasonal variation of daily insolation
'from formulas in Sellers and Berger, 1978.  Uses approximate
'expression for true anomaly
```

```
'Also calculates daily average solar zenith angle for use in SWALBEDO
day = tvar * 365.2422
daytorad = 1 / 365.2422 * 2 * pi    'converts DAY to radiations
omegat = (day - dayp) * daytorad 'mean anomaly measured from perihelion
truan = omegat + ecc * SIN(omegat)    'approximation to true anomaly
rho = (1 - ecc ^ 2) / (1 + ecc * COS(truan))  'r/a
sdecl = SIN(obliq) * SIN(truan + (dayp + 9) * daytorad - pi / 2)'SIN(DECL)
cosdecl = SQR(1 - sdecl * sdecl)
tdecl = sdecl / cosdecl
decl = ATN(tdecl)
FOR jlat = 1 TO nlat
    lat = latmp(jlat)
    cc2s = SIN(lat) * sdecl
    cc2c = COS(lat) * cosdecl
    IF (lat <= decl - pi / 2 OR lat >= decl + pi / 2) THEN
        insol(jlat) = 0
        avecza = 0
    ELSEIF (lat > pi / 2 - decl OR lat < -decl - pi / 2) THEN
        insol(jlat) = solcon / rho ^ 2 * cc2s
        avecza = cc2s
    ELSE
        cc1c = -TAN(lat) * tdecl        'cosH0
        cc1s = SQR(1 - cc1c ^ 2)        'sinH0
        IF cc1c = 0 THEN
            ha0 = pi / 2`
        ELSEIF cc1c < 0 THEN
            ha0 = pi + ATN(cc1s / cc1c)
        ELSE
            ha0 = ATN(cc1s / cc1c)                'Hour angle at sunset
        END IF
    insol(jlat) = solcon / (pi * rho ^ 2)        'Units of watt/m^2
    insol(jlat) = insol(jlat) * (ha0 * cc2s + cc2c * cc1s)
    avecza = cc2s + cc2c * cc1s / ha0    'Average of cosine of zenith angle
    END IF
    IF ABS(avecza) <> 1 THEN
        aveza(jlat) = pi / 2 - ATN(avecza / SQR(1 - avecza * avecza))
    ELSEIF avecza = 1 THEN
        aveza(jlat) = 0
    ELSE
        aveza(jlat) = pi    'Daily average zenith angle
    END IF
    aveza(jlat) = aveza(jlat) * 180 / pi 'Convert radians to degrees
```

```
NEXT jlat
RETURN
'*************************************************************************
Plus subroutine STEPPER from Program DGC04
Subroutine STARTER, STOPPER, FILER, GRINC, PLTC from Program DGC09
Subroutine CHECKSTEP from Program ISOT01
Subroutine SWALBEDO, LWRAD, DEFINITIONS, GRAFINIT, PLOTTER
    from Program DAV07
Subroutine CORE, SLOPERD, and GAUSSD from Program DAV08
```

separate file, as I have done here, and to import this separate file into the spreadsheet either beside or below the temperature.

Results of a calculation with program DAV09 are shown in Figure 7–9 for selected latitudes in the Southern Hemisphere and in Figure 7–10 for selected latitudes in the Northern Hemisphere. The points in these figures are observed zonally averaged temperatures reported by Fairbridge (1967). I tuned the simulation to these data by adjusting the solar constant *solcon*, in subroutine SPECS. I also experimented with different values for the transport coefficient *diffc* and the heat exchange depth in the ocean, *odhc*.

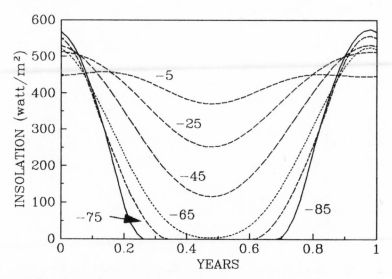

Fig. 7–8. Solar radiation incident on a horizontal surface (insolation) as a function of time at various latitudes in the Southern Hemisphere.

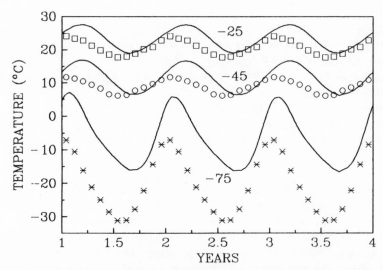

Fig. 7–9. Calculated values of seasonally varying temperatures at three latitudes in the Southern Hemisphere, plotted as solid lines. Observed zonally averaged temperatures at these latitudes are plotted as symbols.

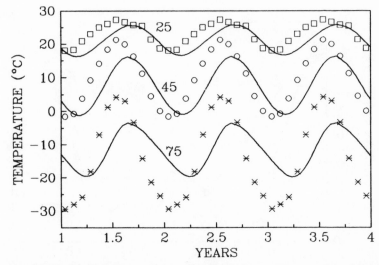

Fig. 7–10. Calculated values of seasonally varying temperatures at three latitudes in the Northern Hemisphere, plotted as solid lines. Observed zonally averaged temperatures at these latitudes are plotted as symbols.

Adjustment of these parameters does not improve the agreement between simulation and observation. The main areas of disagreement are the phase of the simulation, which lags the observation, a simulated temperature at high southern latitudes that is too large, and a simulated amplitude at high northern latitudes that is too small. North, Menzel, and Short (1983) used a transport coefficient that decreases with increasing latitude in order to reduce simulated temperatures at high latitudes. In order to reduce the phase lag between simulation and observations, Thompson and Schneider (1979) used an ocean mixed-layer thickness that varies seasonally. The simulation presented here is very simple, holding both of these coefficients constant. Close agreement between theory and observation is probably not to be expected.

7.6 Sensitivity of the Climate Simulation to Changes in Conditions

It is interesting to explore the response of this simple seasonal climate model to various perturbations. I shall now show just how easy it is to modify the program to see how the simulation is affected by permanent ice in the polar regions, by the suppression of seasonal ice, by an increase in cloud cover, by an increase in atmospheric carbon dioxide, and by continental drift. Although the response of the simulation is not necessarily the same as the response of the real climate system, the simulation nevertheless does conserve energy and does describe insolation, albedo, and long-wave radiation with reasonable care. In addition, it incorporates in a zonal average the marked difference in effective heat capacity between land and sea. If the real climate system does behave in a qualitatively different fashion from the simulation, it is because of factors not included here, factors such as temporally or spatially varying transport coefficient; dependence of cloud cover on something other than land, ocean, and latitude; or an effective heat capacity that varies with time or longitude.

In the study of the response of this simulation to different aspects of the climate system I shall in each case compare the results of the modified system with the results from program DAV09 plotted in Figures 7–9 and 7–10. I shall also compare the albedos of the modified system in January and in June with the albedos calculated by program DAV09. These albedos are plotted in Figure 7–11. The seasonal variation is due to solar zenith angle effects at low latitude augmented, at high latitude, by snow and ice effects described by the temperature-dependent terms in subroutine SWALBEDO.

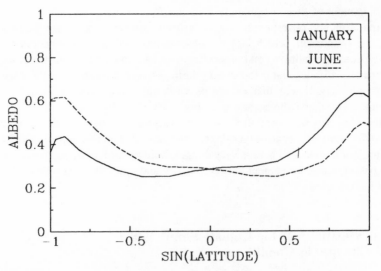

Fig. 7–11. The albedo as a function of the sine of latitude in January and in June.

The differences between the hemispheres are largely the result of different distributions of land as a function of latitude.

7.6.1 The Influence of Permanent Ice

I modify the program to simulate the effect of permanent ice at high latitudes by setting the albedo for the two highest latitudes in each hemisphere equal to 0.7, independent of season, temperature, or solar zenith angle. These values are set at the end of subroutine SWALBEDO. The modified program is listed as DAV10.

The modifications to the albedo are apparent in Figure 7–12, which compares the January and July albedos for DAV10, plotted as solid lines, with those for DAV09, plotted as dashed lines. The only significant difference is at the high latitudes where DAV10 specifies permanent ice with an albedo of 0.7. The influence of this albedo change on temperatures is shown in Figure 7–13 for the Southern Hemisphere and Figure 7–14 for the Northern Hemisphere. In these figures the solid lines refer to results from DAV10, and the dashed lines refer to results from DAV09. Permanent ice reduces summer temperatures in south polar regions by about 10°C. The north polar response is less marked because of the much larger heat capacity of the north polar regions, corresponding to the Arctic Ocean. The

```
'Program DAV10 is an 18 reservoir energy balance climate model
'using the long wave radiation formulation of Kuhn et al. (1989) in LWFLUX
'Albedo formulation of Thompson and Barron in SWALBEDO
'with land and sea ice.
'Seasonal variation of insolation and zenith angle from subroutine SEASON
'This program uses the new solver GAUSSD and SLOPERD
'Albedo in polar reservoirs set to .7 to represent permanent ice
nrow = 18         'the number of equations and unknowns
ncol = nrow + 1
'=====================================================================
'All the dimension statements are collected here
DIM sleq(nrow, ncol), unk(nrow), excoeff(nrow, ncol), ovol(nrow)
DIM y(nrow), dely(nrow), yp(nrow), incind(nrow)
'Establish arrays for graphics routines, GRAFINIT and PLOTTER
numplot = 18         'Number of variables to plot. Not more than 19
DIM ploty(numplot), plotz(numplot), plots(numplot), plotl$(numplot)
nlat = 18         'Radiation subroutines
DIM land(nlat), oceancl(nlat), landcl(nlat), zatemp(nlat)
DIM relarea(nlat)
DIM lwflux(nlat), xlat(nlat), albedo(nlat), insol(nlat)
DIM hcap(nlat), cxlat(nlat), latmp(nlat), aveza(nlat)
'Constants of the seasonal variation of insolation
CONST pi = 3.141593, obliq = 23.45 / 180 * pi, ecc = .017
CONST dayp = 3         'Day number of perihelion
'=====================================================================
GOSUB SPECS
GOSUB FILER            'Choose files for input and output
xstart = 1    'time to start
xend = 4      'time to stop
x = xstart
delx = .001           'measure time in years
mstep = .1
GOSUB CLIMINP
GOSUB CORE
END
'*********************************************************************
SWALBEDO:    'Subroutine calculates the short wave albedo
'approximating the formulation of Thompson and Barron
FOR jlat = 1 TO nlat
'Analytic representation of dependence of albedo on daily average
'zenith angle
    IF y(jlat) < 10 THEN            'Temperature effect
```

```
            allcy = .38 + .0073 * (aveza(jlat) - 50) / (1 + (10 - y(jlat)) / 70
            allcy = allcy + (10 - y(jlat)) / 160
            IF allcy > .7 THEN allcy = .7          'upper limit on land albedo
            allcr = .17 + .0018 * (aveza(jlat) - 50)
            temporary = (1 + (y(jlat) - 10) / 20)
            allcr = allcr + .000065 * (aveza(jlat) - 50) ^ 2 * temporary
            allcr = allcr - .012 * (y(jlat) - 10)
            IF allcr > .7 THEN allcr = .7'upper limit on land albedo
        ELSE
            allcy = .377 + .0075 * (aveza(jlat) - 50)
            allcr = .17 + .0017 * (aveza(jlat) - 50)
            allcr = allcr  + .00009 * (aveza(jlat) - 50) ^ 2
        END IF
        fi = (4 - y(jlat)) / 25       'ice fraction
        IF fi > 1 THEN fi = 1
        IF fi < 0 THEN fi = 0
        alocy = .36 + .0075 * (aveza(jlat) - 50) * (1 - fi / 3.2) + fi * .145
        alocr = .102 + .0002 * (aveza(jlat) - 50)
        alocr = alocr + .00015 * (aveza(jlat) - 50) ^ 2 * (1 - fi / 1.4)
        alocr = alocr + fi * .31
        alland = allcy * landcl(jlat) + allcr * (1 - landcl(jlat))
        alocean = alocy * oceancl(jlat) + alocr * (1 - oceancl(jlat))
        albedo(jlat) = land(jlat) * alland
        albedo(jlat) = albedo(jlat) + (1 - land(jlat)) * alocean
    NEXT jlat
    'Force ice at poles
    albedo(1) = .7
    albedo(2) = .7
    albedo(17) = .7
    albedo(18) = .7
    RETURN
    '****************************************************************************
    Plus subroutine STEPPER from Program DGC04
    Subroutine STARTER, STOPPER, FILER, GRINC, PLTC from Program DGC09
    Subroutine CHECKSTEP from Program ISOT01
    Subroutine LWRAD, DEFINITIONS, GRAFINIT, PLOTTER from Program DAV07
    Subroutine CORE, SLOPERD, and GAUSSD from Program DAV08
    Subroutine SPECS, PRINTER, EQUATIONS, OTHER, CLIMINP,
        and SEASON from Program DAV09
```

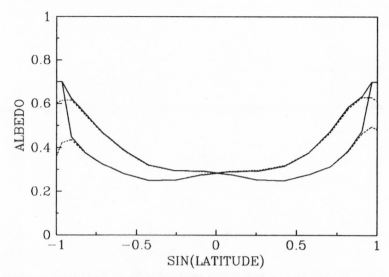

Fig. 7–12. The albedo as a function of the sine of latitude in January and in June. The solid lines are the albedo from program DAV10; the dashed lines are from program DAV09.

effect on winter temperatures is not as large because albedo does not matter during the polar winter when the insolation is zero. The response of lower-latitude temperatures to permanent polar ice is small, a few degrees at 45° south latitude and less at the other latitudes plotted. Permanent polar ice therefore increases the temperature difference between the middle latitudes and the poles, particularly in summer. But because the polar latitudes represent a small fraction of the area of the globe, the impact on the global average temperature of permanent polar ice is small.

7.6.2 The Influence of Seasonal Ice

As an alternative numerical experiment, I suppress all ice formation by modifying subroutine SWALBEDO. The modifications are to set the fraction of the ocean covered by sea ice to zero, regardless of temperature, and to set an unattainably low temperature limit in the IF statement that branches to temperature-dependent albedo on land. The program is listed as DAV11.

The influence of these changes can be seen in Figure 7–15, which plots the January and July albedo values. The solid lines here correspond to

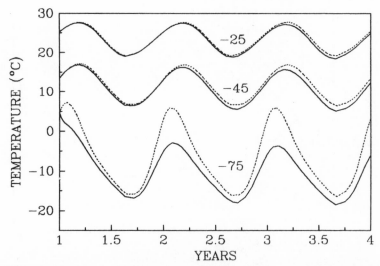

Fig. 7–13. The seasonal variation of temperature in the Southern Hemisphere plotted as solid lines for DAV10, with permanent ice at high latitudes, and as dashed lines for the reference results, DAV09. The new calculations are started with the same initial values. It takes a few years for the influence of the initial values to die out.

DAV11, the dashed lines to DAV09. The calculation, like all the calculations in this series of numerical experiments, is begun at the same initial values as is DAV09. Calculated temperatures are plotted in Figures 7–16 and 7–17, with the solid lines corresponding to DAV11 and the dashed lines corresponding to DAV09. Because the suppression of ice formation reduces the albedo, as shown in Figure 7–15, the new temperatures are everywhere larger than the reference temperatures. The change is most marked in the Northern Hemisphere, at high latitudes, where the temperature increase amounts to about 5°C in the summertime. The greater sensitivity of the Northern Hemisphere is probably a consequence of a higher fraction of land in middle to high latitudes. As Figure 7–3 shows, the sensitivity of albedo to temperature is greater for land than for ocean. In addition, temperatures at high latitudes in the Northern Hemisphere are low enough to permit ice nearly all year round in the calculations of DAV09. On the other hand, temperatures in the south polar region fall below freezing mainly in the winter months, when albedo is relatively unimportant.

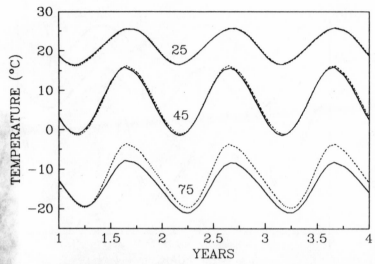

Fig. 7–14. The seasonal variation of temperature in the Northern Hemisphere plotted as solid lines for DAV10, with permanent ice at high latitudes, and as dashed lines for the reference results, DAV09. The new calculations are started with the same initial values. It takes a few years for the influence of the initial values to die out.

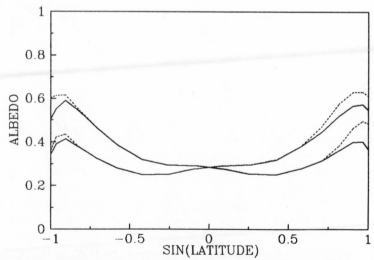

Fig. 7–15. Albedos in January and in July plotted as solid lines for DAV11, with no ice and snow, and as dashed lines for the reference results, DAV09.

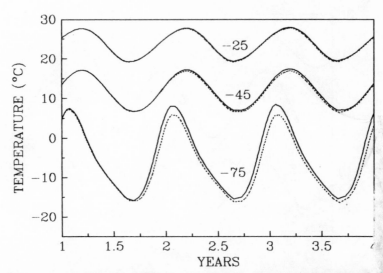

Fig. 7–16. The seasonal variation of temperature in the Southern Hemisp. plotted as solid lines for DAV11, with no ice and snow, and as dashed lines for reference results, DAV09.

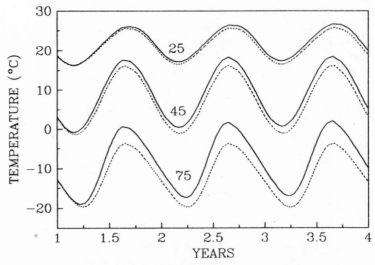

Fig. 7–17. The seasonal variation of temperature in the Northern Hemisphere plotted as solid lines for DAV11, with no ice and snow, and as dashed lines for the reference results, DAV09.

```
'Program DAV11 is an 18 reservoir energy balance climate model
'using the long wave radiation formulation of Kuhn et al. (1989) in LWFLUX
'Albedo formulation of Thompson and Barron in SWALBEDO
'with land and sea ice.
'Seasonal variation of insolation and zenith angle from subroutine SEASON
'This program uses the new solver GAUSSD and SLOPERD
'All ice is suppressed in SWALBEDO
nrow = 18        'the number of equations and unknowns
ncol = nrow + 1
'=======================================================================
'All the dimension statements are collected here
DIM sleq(nrow, ncol), unk(nrow), excoeff(nrow, ncol), ovol(nrow)
DIM y(nrow), dely(nrow), yp(nrow), incind(nrow)
'Establish arrays for graphics routines, GRAFINIT and PLOTTER
numplot = 18        'Number of variables to plot. Not more than 19
DIM ploty(numplot), plotz(numplot), plots(numplot), plotl$(numplot)
nlat = 18  ·        'Radiation subroutines
DIM land(nlat), oceancl(nlat), landcl(nlat), zatemp(nlat)
DIM relarea(nlat)
DIM lwflux(nlat), xlat(nlat), albedo(nlat), insol(nlat)
DIM hcap(nlat), cxlat(nlat), latmp(nlat), aveza(nlat)
'Constants of the seasonal variation of insolation
CONST pi = 3.141593, obliq = 23.45 / 180 * pi, ecc = .017
CONST dayp = 3        'Day number of perihelion
'=======================================================================
GOSUB SPECS
GOSUB FILER             'Choose files for input and output
xstart = 1     'time to start
xend = 4       'time to stop
x = xstart
delx = .001        'measure time in years
mstep = .1
GOSUB CLIMINP
GOSUB CORE
END
'***********************************************************************
SWALBEDO:    'Subroutine calculates the short wave albedo.
FOR jlat = 1 TO nlat
```

```
'Analytic representation of dependence of albedo on daily average
'zenith angle
    IF y(jlat) < -100 THEN         'Temperature effect suppressed
        allcy = .38 + .0073 * (aveza(jlat) - 50) / (1 + (10 - y(jlat)) / 70)
        allcy = allcy + (10 - y(jlat)) / 160
        IF allcy > .7 THEN allcy = .7          'upper limit on land albedo
        allcr = .17 + .0018 * (aveza(jlat) - 50)
        temporary = (1 + (y(jlat) - 10) / 20)
        allcr = allcr + .000065 * (aveza(jlat) - 50) ^ 2 * temporary
        allcr = allcr - .012 * (y(jlat) - 10)
        IF allcr > .7 THEN allcr = .7'upper limit on land albedo
    ELSE
        allcy = .377 + .0075 * (aveza(jlat) - 50)
        allcr = .17 + .0017 * (aveza(jlat) - 50)
        allcr = allcr  + .00009 * (aveza(jlat) - 50) ^ 2
    END IF
    fi = 0   '(4 - y(jlat)) / 25      'ice fraction set to 0
    IF fi > 1 THEN fi = 1
    IF fi < 0 THEN fi = 0
    alocy = .36 + .0075 * (aveza(jlat) - 50) * (1 - fi / 3.2) + fi * .145
    alocr = .102 + .0002 * (aveza(jlat) - 50)
    alocr = alocr + .00015 * (aveza(jlat) - 50) ^ 2 * (1 - fi / 1.4)
    alocr = alocr + fi * .31
    alland = allcy * landcl(jlat) + allcr * (1 - landcl(jlat))
    alocean = alocy * oceancl(jlat) + alocr * (1 - oceancl(jlat))
    albedo(jlat) = land(jlat) * alland
    albedo(jlat) = albedo(jlat) + (1 - land(jlat)) * alocean
NEXT jlat
RETURN
'*********************************************************************************
Plus subroutine STEPPER from Program DGC04
Subroutine STARTER, STOPPER, FILER, GRINC, PLTC from Program DGC09
Subroutine CHECKSTEP from Program ISOT01
Subroutine LWRAD, DEFINITIONS, GRAFINIT, PLOTTER from Program DAV07
Subroutine CORE, SLOPERD, and GAUSSD from Program DAV08
Subroutine SPECS, PRINTER, EQUATIONS, OTHER, CLIMINP,
    and SEASON from Program DAV09
```

7.6.3 The Influence of Clouds

As an alternative experiment I increase cloud cover uniformly by a factor of 1.1 at all latitudes, both on land and at sea. The modification is made in subroutine CLIMINP, which reads the values of cloud cover from a file. The modified program is listed as DAV12.

Figure 7–18 shows the relatively uniform increase in albedo at all latitudes and seasons that results from the increase in cloud cover. The temperature results are plotted in Figures 7–19 and 7–20. Increased cloud cover results in a decrease of temperature by not more than a degree or so at most latitudes and seasons. Simulated temperatures are remarkably insensitive to cloud cover because the effect of clouds on albedo is counteracted by the effect of clouds on outgoing long-wave radiation. Increased cloud cover decreases solar energy absorbed by the Earth, and it also decreases the outgoing flux of long-wave radiation, as shown in Figure 7–5, because the outgoing radiation in this formulation leaves from the cold cloud tops. The results in Figures 7–19 and 7-20 show, indeed, that the cancellation of these two effects is nearly perfect in the polar regions in wintertime. At lower latitudes, where albedo is more important, the influ-

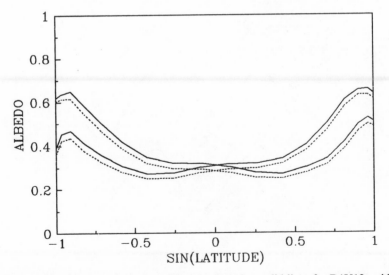

Fig. 7–18. Albedos in January and in July plotted as solid lines for DAV12, with cloud cover increased by a factor of 1.1, and as dashed lines for the reference program, DAV09.

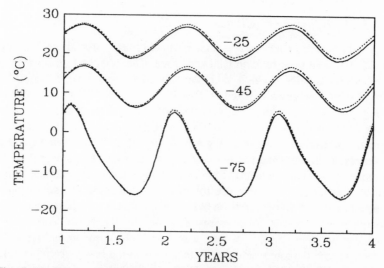

Fig. 7–19. The seasonal variation of temperature in the Southern Hemisphere plotted as solid lines for DAV12, with cloud cover increased by a factor of 1.1, and as dashed lines for the reference results, DAV09.

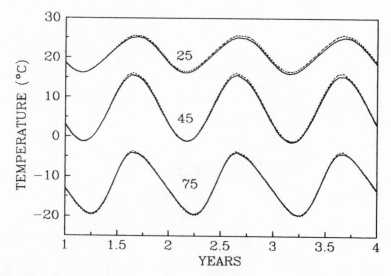

Fig. 7–20. The seasonal variation of temperature in the Northern Hemisphere plotted as solid lines for DAV12, with cloud cover increased by a factor of 1.1, and as dashed lines for the reference results, DAV09.

```
'Program DAV12 is an 18 reservoir energy balance climate model
'using the long wave radiation formulation of Kuhn et al. (1989) in LWFLUX
'Albedo formulation of Thompson and Barron in SWALBEDO
'with land and sea ice.
'Seasonal variation of insolation and zenith angle from subroutine SEASON
'This program uses the new solver GAUSSD and SLOPERD
'Cloud cover is increased by factor of 1.1 in CLIMINP
nrow = 18        'the number of equations and unknowns
ncol = nrow + 1
'======================================================================
'All the dimension statements are collected here
DIM sleq(nrow, ncol), unk(nrow), excoeff(nrow, ncol), ovol(nrow)
DIM y(nrow), dely(nrow), yp(nrow), incind(nrow)
'Establish arrays for graphics routines, GRAFINIT and PLOTTER
numplot = 18        'Number of variables to plot. Not more than 19
DIM ploty(numplot), plotz(numplot), plots(numplot), plotl$(numplot)
nlat = 18  ·        'Radiation subroutines
DIM land(nlat), oceancl(nlat), landcl(nlat), zatemp(nlat)
DIM relarea(nlat)
DIM lwflux(nlat), xlat(nlat), albedo(nlat), insol(nlat)
DIM hcap(nlat), cxlat(nlat), latmp(nlat), aveza(nlat)
'Constants of the seasonal variation of insolation
CONST pi = 3.141593, obliq = 23.45 / 180 * pi, ecc = .017
CONST dayp = 3        'Day number of perihelion
'======================================================================
GOSUB SPECS
GOSUB FILER            'Choose files for input and output
xstart = 1      'time to start
xend = 4      'time to stop
x = xstart
delx = .001          'measure time in years
mstep = .1
GOSUB CLIMINP
GOSUB CORE
END
'**********************************************************************
CLIMINP:    'Subroutine reads land area, sin(latitude), and other
'parameters of the climate system
OPEN "CLIMINP.PRN" FOR INPUT AS #5
FOR jvar = 1 TO nlat
    INPUT #5, latmp(jvar), xlat(jvar), relarea(jvar), land(jvar)
    INPUT #5, landcl(jvar), oceancl(jvar), insol(jvar), aveza(jvar)
```

```
NEXT jvar
CLOSE #5
FOR jlat = 1 TO nlat
    hcap(jlat) = hcconst * ((1 - land(jlat)) + land(jlat) / hcrat)
    'Assume heat capacity of land is 1/hcrat times that of ocean
    cxlat(jlat) = SQR(1 - xlat(jlat) ^ 2)
    latmp(jlat) = latmp(jlat) / 180 * pi        'latitude in radians
    landcl(jlat) = landcl(jlat) * 1.1           'increase cloud cover
    oceancl(jlat) = oceancl(jlat) * 1.1
NEXT jlat
'Output seasonally varying parameters
OPEN "RESSEAS.PRN" FOR OUTPUT AS #5
RETURN
'**************************************************************************
Plus subroutine STEPPER from Program DGC04
Subroutine STARTER, STOPPER, FILER, GRINC, PLTC from Program DGC09
Subroutine CHECKSTEP from Program ISOT01
Subroutine SWALBEDO, LWRAD, DEFINITIONS, GRAFINIT, PLOTTER
    from Program DAV07
Subroutine CORE, SLOPERD, and GAUSSD from Program DAV08
Subroutine SPECS, PRINTER, EQUATIONS, OTHER, and SEASON from Program DAV09
```

ence on albedo has a little more impact than does the influence on outgoing planetary radiation.

7.6.4 The Influence of Carbon Dioxide

This program can explore the influence of change in the carbon dioxide amount simply by changing the value of *pco2*, set in subroutine SPECS. Here I set atmospheric carbon dioxide to four times the present level, producing program DAV13.

Figure 7–21 shows that the increased greenhouse effect results in less ice and snow at high latitudes and a small reduction in the high-latitude albedos in January and July. The response of temperatures is shown in Figures 7–22 and 7–23: higher temperatures at all latitudes and seasons, a change of about 3° in this simulation. The parameters used in subroutine LWFLUX result in a relatively small response to modest increases in carbon dioxide; other simulations are more sensitive to carbon dioxide change. The magnitude of the temperature response depends little on lati-

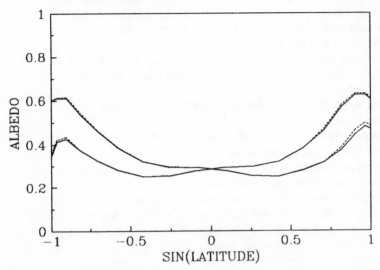

Fig. 7–21. The albedo in January and in July plotted as solid lines for DAV13, with atmospheric carbon dioxide increased by a factor of 4, and as dashed lines for the reference results, DAV09.

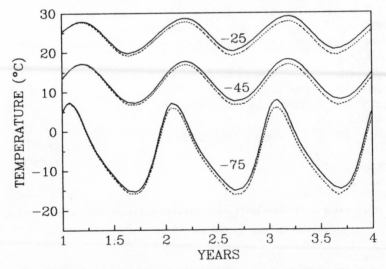

Fig. 7–22. The seasonal variation of temperature in the Southern Hemisphere plotted as solid lines for DAV13, with atmospheric carbon dioxide increased by a factor of 4, and as dashed lines for the reference results, DAV09.

```
'Program DAV13 is an 18 reservoir energy balance climate model
'using the long wave radiation formulation of Kuhn et al. (1989) in LWFLUX
'Albedo formulation of Thompson and Barron in SWALBEDO
'with land and sea ice.
'Seasonal variation of insolation and zenith angle from subroutine SEASON
'This program uses the new solver GAUSSD and SLOPERD
'PCO2=4 set in SPECS
nrow = 18        'the number of equations and unknowns
ncol = nrow + 1
'=======================================================================
'All the dimension statements are collected here
DIM sleq(nrow, ncol), unk(nrow), excoeff(nrow, ncol), ovol(nrow)
DIM y(nrow), dely(nrow), yp(nrow), incind(nrow)
'Establish arrays for graphics routines, GRAFINIT and PLOTTER
numplot = 18        'Number of variables to plot. Not more than 19
DIM ploty(numplot), plotz(numplot), plots(numplot), plotl$(numplot)
nlat = 18 ·        'Radiation subroutines
DIM land(nlat), oceancl(nlat), landcl(nlat), zatemp(nlat)
DIM relarea(nlat)
DIM lwflux(nlat), xlat(nlat), albedo(nlat), insol(nlat)
DIM hcap(nlat), cxlat(nlat), latmp(nlat), aveza(nlat)
'Constants of the seasonal variation of insolation
CONST pi = 3.141593, obliq = 23.45 / 180 * pi, ecc = .017
CONST dayp = 3        'Day number of perihelion
'=======================================================================
GOSUB SPECS
GOSUB FILER           'Choose files for input and output
xstart = 1    'time to start
xend = 4      'time to stop
x = xstart
delx = .001           'measure time in years
mstep = .1
GOSUB CLIMINP
GOSUB CORE
END
'***********************************************************************
SPECS: 'Subroutine to read in the specifications of the problem
pco2 = 4
solcon = 1380        'W/m^2
```

```
diffc = 3          'Heat transport
odhc = 35          'Heat exchange depth in ocean in meters
hcrat = 50         'Heat constant ratio, ocean to land
hcconst = 4185500! * odhc
secpy = 365.25 * 24 * 3600
'Temperature in degrees C.  Starting values from Sellers
y(18) = -23.4
y(17) = -15.7
y(16) = -7
y(15) = .7
y(14) = 7.7
y(13) = 14.2
y(12) = 20.6
y(11) = 25.3
y(10) = 25.7
y(9) = 25
y(8) = 23.5
y(7) = 19
y(6) = 13.7
y(5) = 8.9
y(4) = 1.4
y(3) = -10.8
y(2) = -19.3
y(1) = -47.7
'INCIND=1 for test on relative increment
FOR jlat = 1 TO nlat
    incind(jlat) = 3     'Absolute increment < 3 degrees
NEXT jlat
RETURN
'****************************************************************************
Plus subroutine STEPPER from Program DGC04
Subroutine STARTER, STOPPER, FILER, GRINC, PLTC from Program DGC09
Subroutine CHECKSTEP from Program ISOT01
Subroutine SWALBEDO, LWRAD, DEFINITIONS, GRAFINIT, PLOTTER
    from Program DAV07
Subroutine CORE, SLOPERD, and GAUSSD from Program DAV08
Subroutine PRINTER, EQUATIONS, OTHER, CLIMINP, and SEASON
    from Program DAV09
```

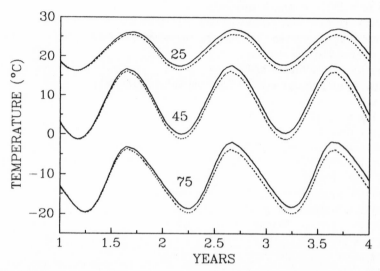

Fig. 7–23. The seasonal variation of temperature in the Northern Hemisphere plotted as solid lines for DAV13, with atmospheric carbon dioxide increased by a factor of 4, and as dashed lines for the reference results, DAV09.

tude or season. In this simulation, increased carbon dioxide warms the globe uniformly. The frequently quoted assertion that the carbon dioxide greenhouse will warm high latitudes more than low ones must depend on aspects of the climate system not included here, such as changing cloud cover or changing heat transport. At least in this simulation, the impact of the greenhouse effect on albedo through high-latitude snow and ice is not large enough to cause markedly greater temperature increases at high latitudes than at low ones.

7.6.5 The Influence of Geography

As a final application of the seasonal climate simulation, I consider the influence of continental drift. The input file CLIMINP.PRN is modified to contain the land distribution for the Miocene, 15 million years ago, as tabulated by Parrish (1985). The program, DAV14, is otherwise unmodified.

The Miocene land distribution is compared with the present land distribution in Figure 7–24. There has been an increase in land fraction at the South Pole since the Miocene, a decrease in middle latitudes in the South-

```
'Program DAV14 is an 18 reservoir energy balance climate model
'using the long wave radiation formulation of Kuhn et al. (1989) in LWFLUX
'Albedo formulation of Thompson and Barron in SWALBEDO
'with land and sea ice.
'Seasonal variation of insolation and zenith angle from subroutine SEASON
'This program uses the new solver GAUSSD and SLOPERD
'Miocene land distribution from Parrish entered in file CLIMINP.PRN
nrow = 18        'the number of equations and unknowns
ncol = nrow + 1
'=======================================================================
'All the dimension statements are collected here
DIM sleq(nrow, ncol), unk(nrow), excoeff(nrow, ncol), ovol(nrow)
DIM y(nrow), dely(nrow), yp(nrow), incind(nrow)
'Establish arrays for graphics routines, GRAFINIT and PLOTTER
numplot = 18       'Number of variables to plot.  Not more than 19
DIM ploty(numplot), plotz(numplot), plots(numplot), plotl$(numplot)
nlat = 18  ·       'Radiation subroutines
DIM land(nlat), oceancl(nlat), landcl(nlat), zatemp(nlat)
DIM relarea(nlat)
DIM lwflux(nlat), xlat(nlat), albedo(nlat), insol(nlat)
DIM hcap(nlat), cxlat(nlat), latmp(nlat), aveza(nlat)
'Constants of the seasonal variation of insolation
CONST pi = 3.141593, obliq = 23.45 / 180 * pi, ecc = .017
CONST dayp = 3     'Day number of perihelion
'=======================================================================
GOSUB SPECS
GOSUB FILER            'Choose files for input and output
xstart = 1     'time to start
xend = 4     'time to stop
x = xstart
delx = .001        'measure time in years
mstep = .1
GOSUB CLIMINP         'File CLIMINP.PRN has been modified
GOSUB CORE
END
'***********************************************************************
Plus subroutine STEPPER from Program DGC04
Subroutine STARTER, STOPPER, FILER, GRINC, PLTC from Program DGC09
Subroutine CHECKSTEP from Program ISOT01
Subroutine SWALBEDO, LWRAD, DEFINITIONS, GRAFINIT, PLOTTER
    from Program DAV07
```

Subroutine CORE, SLOPERD, and GAUSSD from Program DAV08
Subroutine SPECS, PRINTER, EQUATIONS, OTHER, CLIMINP,
    and SEASON from Program DAV09

ern Hemisphere, an increase in land fraction at middle latitudes in the Northern Hemisphere, and a decrease in land fraction in northern polar regions. The influence of these changes on albedo is small at most latitudes, as shown in Figure 7–25. The only change that shows on these plots is in the winter polar regions, which is a consequence of changed temperatures and the albedo of snow and ice. At lower latitudes the changes in land fraction and cloud cover are too small to influence the albedo significantly.

The calculated temperatures are plotted in Figures 7–26 and 7–27, with the changes at the middle and low latitudes in the Southern Hemisphere being quite small. The seasonal amplitude in the south polar region has increased from the Miocene to the present, however, because of the increase in land fraction. Changes in the Northern Hemisphere are the opposite, with the seasonal amplitude having increased slightly at middle and low latitudes because land fraction has increased from the Miocene to the

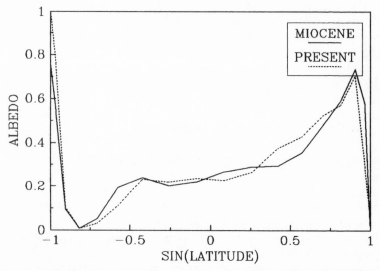

Fig. 7–24. Land fraction as a function of the sine of latitude in the Miocene, 15 million years ago, plotted as a solid line, and at present, plotted as a dashed line.

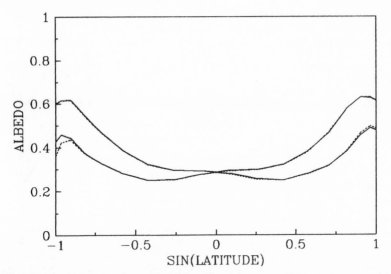

Fig. 7–25. Albedos in January and in July plotted as solid lines for DAV14, with Miocene land distribution, and as dashed lines for the present land distribution, DAV09.

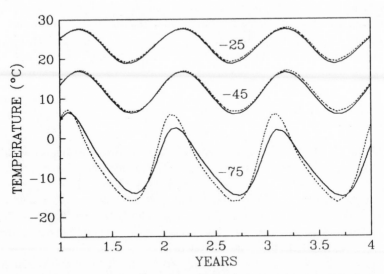

Fig. 7–26. The seasonal variation of temperature in the Southern Hemisphere plotted as solid lines for the Miocene land distribution, DAV14, and as dashed lines for the present land distribution, DAV09.

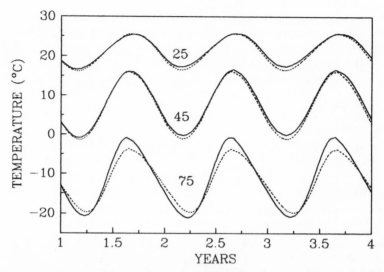

Fig. 7–27. The seasonal variation of temperature in the Northern Hemisphere plotted as solid lines for the Miocene land distribution, DAV14, and as dashed lines for the present land distribution, DAV09.

present. On the other hand, the seasonal amplitude has decreased in the polar regions because of the decrease in land fraction. The overall effect is a reduction in maximum temperatures in the Arctic region, which might have played a role in the onset of Pleistocene glaciation in the Northern Hemisphere.

7.7 Summary

This chapter demonstrated the computational simplification that is possible in systems consisting of a one-dimensional chain of coupled reservoirs, which arise in diffusion and heat conduction problems. In such systems each equation is coupled just to its immediate neighbors, so that much of the work involved in Gaussian elimination and back substitution can be avoided. I presented here two subroutines, GAUSSD and SLOPERD, that deal efficiently with this kind of system.

I also applied the revised computational method to calculate zonally averaged temperature as a function of latitude. I introduced an energy balance climate model, which calculates surface temperature for absorbed solar energy, emitted planetary radiation, and the transport of heat between latitudes by winds and ocean currents. Much of the work of the program is describing insolation, albedo, and long-wave radiation as functions of position, time, and temperature. The simulation was developed first in an annually averaged version for which the specification of incident solar energy is straightforward. I then considered a more interesting seasonal simulation, which includes a calculation of insolation as a function of latitude and time. This simulation is used to explore the influence on simulated temperatures of permanent ice at high latitudes, seasonal ice at high latitudes, changing cloud cover, increased carbon dioxide amount, and continental drift.

These applications show how easy it is to modify a program, once it has been developed, to explore such questions. Another application, not undertaken here, would be to explore Milankovitch perturbations in the Earth's orbital parameters, eccentricity, obliquity, and date of perihelion. These parameters are specified as constants at the beginning of the program, and it would be simple to change their values as predicted by astronomical calculations, in order to see how the seasonal variation of temperature is affected at various latitudes.

8 Interacting Species
in Identical Reservoirs

8.1 Introduction

In Chapter 7 I showed how much computational effort could be avoided in a system consisting of a chain of identical equations each coupled just to its neighboring equations. Such systems arise in linear diffusion and heat conduction problems. It is possible to save computational effort because the *sleq* array that describes the system of simultaneous linear algebraic equations that must be solved has elements different from zero on and immediately adjacent to the diagonal only.

This general approach works also for one-dimensional diffusion problems involving several interacting species. In such a system the concentration of a particular species in a particular reservoir is coupled to the concentrations of other species in the same reservoir by reactions between species and is coupled also to adjacent reservoirs by transport between reservoirs. If the differential equations that describe such a system are arranged in appropriate order, with the equations for each species in a given reservoir followed by the equations for each species in the next reservoir and so on, the *sleq* array still will have elements different from zero close to the diagonal only. All the nonzero elements lie no farther from the diagonal than the number of species. More distant elements are zero. Again, much computation can be eliminated by taking advantage of this pattern. I will show how to solve such a system in this chapter, introducing two new solution subroutines, GAUSSND and SLOPERND, to replace GAUSSD and SLOPERD.

I shall apply the new method of solution to a problem of early diagenesis in carbonate sediments. I calculate the properties of the pore fluid in the sediment as a function of depth and time. The different reservoirs are successive layers of sediment at increasing depth. The fluid's composition is affected by diffusion between sedimentary layers and between the top layer and the overlying seawater, the oxidation of organic carbon, and the dissolution or precipitation of calcium carbonate.

Because I assume that the rate of oxidation of organic carbon decreases exponentially with increasing depth, there must be more chemical activity at shallow depths in the sediment than at great depths. The use of sedimentary layers of uniform thickness is not optimal for this situation, and so I shall show how much the solution is improved by using thin layers at shallow depths and thicker layers at greater depths. I then introduce the new solver, which speeds up the calculation significantly by ignoring the elements in the *sleq* array that are far from the diagonal and are zero. The solution now uses variable layer thickness and avoids unnecessary computation. I take advantage of the efficiency of this simulation to add equations for the stable carbon isotope ratio in the pore fluids, a further complication of the system that is easily accommodated.

8.2 Diagenesis in Carbonate Sediments

I consider a system in which organic matter is oxidized at a steady rate that is a specified function of depth in uniform calcium carbonate sediments. The oxidation of organic matter increases the total dissolved carbon in the pore water of the sediment. The resultant increase in acidity causes the dissolution of calcium carbonate and a consequent increase in alkalinity as well as another increase in total dissolved carbon. The total dissolved carbon and alkalinity are transported by diffusion between different depths in the sediment.

In program SEDS01 I divide the sediment into ten layers each 30 centimeters thick. Parameters are specified in subroutine SPECS: The parameter *dsed* is the diffusion coefficient; *sedl* is the layer thickness; and *prsty* is the porosity of the sediments. The organic carbon oxidation rate is specified in the array *resp*, which I take to be an exponentially decreasing function of depth with a value equal to *resp0* at the sediment–water interface and with a decay constant of *respdec*. The sediments exchange dissolved constituents with seawater at the sediment–water interface. The concentrations in seawater are specified by *casw* for calcium ions, *sigcs* for total dissolved carbon, and *alks* for alkalinity. For the rate of dissolution of calcium carbonate I use the expression introduced in Chapter 6, which expresses dissolution rate *diss* as the difference of exponential functions of

```
'Program SEDS01 solves diagenesis in carbonate sediments
'Equal layer thicknesses
numlay = 10                    'number of layers of sediment
nrow = 20                'the number of equations and unknowns
ncol = nrow + 1
'====================================================================
DIM sleq(nrow, ncol), unk(nrow), y(nrow), dely(nrow), yp(nrow)
DIM incind(nrow)
'Establish graphics arrays
numplot = 19              'Number of variables to plot.  Not more than 19
DIM plotz(numplot), plots(numplot), plotl$(numplot), ploty(numplot)
DIM resp(numlay), diss(numlay)          'sedimentary rates
'====================================================================
GOSUB SPECS
GOSUB FILER
xstart = 1               'time to start
xend = 1500              'time to stop, in years
x = xstart
delx = 1
mstep = 100
GOSUB CORE
END
'*******************************************************************************
PRINTER:   'Subroutine writes a file for subsequent plotting
GOSUB OTHER                      'print current values
PRINT #1, sigcs;
FOR jrow = 1 TO numlay      'Odd numbers are total carbon
    PRINT #1, y(2 * jrow - 1);
NEXT jrow
PRINT #1, casw;
FOR jrow = 1 TO numlay      'Even numbers are calcium
    PRINT #1, y(2 * jrow);
NEXT jrow
PRINT #1, x
GOSUB PLOTTER
RETURN
'*******************************************************************************
EQUATIONS: 'The differential equations that define the simulation
GOSUB OTHER
'Odd numbers are total dissolved carbon, even are calcium
yp(1) = ((sigcs - y(1)) * 2 + (y(3) - y(1))) * transcon + resp(1) + diss(1)
yp(2) = ((casw - y(2)) * 2 + (y(4) - y(2))) * transcon + diss(1)
```

```
FOR jlay = 2 TO numlay - 1
    jsc = 2 * jlay - 1: jca = 2 * jlay
    yp(jsc) = (y(jsc - 2) - 2 * y(jsc) + y(jsc + 2)) * transcon
    yp(jsc) = yp(jsc) + resp(jlay) + diss(jlay)       'continuation
    yp(jca) = (y(jca - 2) - 2 * y(jca) + y(jca + 2)) * transcon
    yp(jca) = yp(jca) + diss(jlay)       'continuation
NEXT jlay
jsc = 2 * numlay - 1: jca = 2 * numlay
yp(jsc) = (y(jsc - 2) - y(jsc)) * transcon + resp(numlay) + diss(numlay)
yp(jca) = (y(jca - 2) - y(jca)) * transcon + diss(numlay)
RETURN
'****************************************************************************
OTHER: 'Subroutine evaluates quantities that change with time but that
'are not dependent variables
FOR jlay = 1 TO numlay
    'Carbonate equilibria
    sigc = y(2 * jlay - 1): alk = alks + 2 * (y(2 * jlay) - casw)
    GOSUB CARBONATE
    omegadel = co3 * y(2 * jlay) / csat - 1
    'carbonate dissolution rate
    diss(jlay) = (EXP(-disscon * omegadel) - EXP(pcpcon * omegadel))
    diss(jlay) = diss(jlay) * disfac
NEXT jlay
RETURN
'****************************************************************************
SPECS: 'Subroutine to read in the specifications of the problem
secpy = 365.25 * 3600 * 24    'seconds per year
difsed = .000002            'cm^2/sec.  Broecker and Peng, p. 55
dsed = difsed * secpy       'cm^2/y
sedl = 30                   'thickness of layers in cm
prsty = .5                  'porosity
transcon = difsed / (sedl ^ 2 * prsty)    'effective diffusion coefficient
respdec = 50                'respiration decay depth in cm
resp0 = .000001             'respiration rate in mole/cm^3/y
resp0 = resp0 * 1000000!'mole/m^3/y
FOR jlay = 1 TO numlay
    'respiration rate
    resp(jlay) = resp0 * EXP(-(jlay - .5) * sedl / respdec)
NEXT jlay
casw = 10                   'calcium ions in surface water, mole/m^3
csat = .46
'calcite saturation constant in mole^2/m^6, Broecker and Peng, p. 59
```

```
disscon = 7        'dissolution constant in DISS(JLAY)
pcpcon = 1         'carbonate precipitation constant
disfac = .01       'scaling factor in dissolution rate mole/m^3/y
sigcs = 2.01       'mole/m3
alks = 2.2
watemp = 288       'temperature of surface water, deg K
FOR jlay = 1 TO numlay            'initial values equal sea water values
    y(2 * jlay - 1) = sigcs
    y(2 * jlay) = casw
NEXT jlay
FOR jrow = 1 TO nrow
    incind(jrow) = 1
NEXT jrow
RETURN
'****************************************************************************
GRAFINIT:                  'Initialize graphics
'Plot relative departures from PLOTZ
'Sensitivity of the plot depends on PLOTS
'The labels are PLOTL$
FOR jplot = 1 TO numplot
    plotz(jplot) = y(jplot) 'Departures from sea water
    plots(jplot) = y(jplot) / 5
    plotl$(jplot) = STR$(jplot)
NEXT jplot
GOSUB GRINC
RETURN
END
'****************************************************************************
PLOTTER:          'Subroutine plots values as the calculation proceeds
'The scale is normalized relative to the starting values
FOR jplot = 1 TO numplot          'Which values to plot
    ploty(jplot) = y(jplot)
NEXT jplot
GOSUB PLTC
RETURN
END
'****************************************************************************
CORE:    'Subroutine directs the calculation
maxinc = .1
```

```basic
    dlny = .001
    count = 0                   'count time steps
    GOSUB GRAFINIT
    GOSUB PRINTER
    DO WHILE x < xend
        count = count + 1 + 1
        GOSUB SLOPER
        GOSUB GAUSS
        FOR jrow = 1 TO nrow
            dely(jrow) = unk(jrow)
        NEXT jrow
        GOSUB CHECKSTEP
    LOOP
    '=========================================================================
    sigc = sigcs: alk = alks           'Print carbonate parameters
    GOSUB CARBONATE
    omegadel = co3 * casw / csat - 1
    'carbonate dissolution rate
    dissw = (EXP(-disscon * omegadel) - EXP(pcpcon * omegadel))
    diss(jlay) = diss(jlay) * disfac
    PRINT #1, omegadel; dissw
    FOR jlay = 1 TO numlay
    'Carbonate equilibria
        sigc = y(2 * jlay - 1): alk = alks + 2 * (y(2 * jlay) - casw)
        GOSUB CARBONATE
        omegadel = co3 * y(2 * jlay) / csat - 1
        'carbonate dissolution rate
        diss(jlay) = (EXP(-disscon * omegadel) - EXP(pcpcon * omegadel)) * disfac
        PRINT #1, omegadel; diss(jlay); resp(jlay)
    NEXT jlay
    GOSUB STOPPER
    RETURN
    '*************************************************************************
    Plus subroutines GAUSS and SWAPPER from Program DGC03
    Subroutine STEPPER from Program DGC04
    Subroutine SLOPER from Program DGC08
    Subroutine CARBONATE, GRINC, PLTC, FILER, STOPPER, and STARTER
        from Program DGC09
    Subroutine CHECKSTEP from Program ISOT01
```

the saturation index, *omegadel*. The starting values of total dissolved carbon and calcium ions are set equal to the seawater values at all depths.

I solve differential equations for the two constituents, total dissolved carbon and calcium ions. These equations are arranged with odd-numbered equations referring to total dissolved carbon and even-numbered equations referring to calcium ions. The total dissolved carbon is affected by diffusion between adjacent layers, by respiration at the rate *resp,* and by dissolution of calcium carbonate at the rate *diss.* The respiration rate is fixed and is specified in subroutine SPECS. The dissolution rate, which may be positive or negative depending on the saturation state of the pore water, is calculated in subroutine OTHER. The transport parameter, *transcon,* is specified in subroutine SPECS; it combines the diffusion coefficient, the thickness of the layers, and the porosity of the sediments. The equations for calcium concentration are like the equations for total dissolved carbon except that they lack the respiration source. I assume that the only process affecting alkalinity is the change in calcium ion concentration. Alkalinity is therefore calculated from seawater alkalinity and calcium concentration by an algebraic expression in subroutine OTHER. I have added to subroutine CORE a section of code that calculates the parameters of the carbon system for the final depth profile of concentrations and prints out these parameters: saturation index, dissolution rate, and respiration rate. My intention is to run the calculation long enough to achieve a steady state and then to examine the variation of these parameters with depth in the sediment.

Figure 8–1 shows the evolution of the depth profile of total dissolved carbon. The numbers labeling the curves in this figure are times from the start of the calculation, expressed in years. The concentration at first increases most rapidly at the shallower depths, where the respiration rate is the largest. After 1600 years the solution is close to steady state; the concentration has risen to about 30 moles per cubic meter and is approximately independent of depth below about 200 centimeters. The 30-centimeter-thick layers used for this calculation show up in Figure 8–1 as straight-line segments in the concentration profile. Figure 8–2 suggests that this calculation requires finer resolution in depth.

Figure 8–2 shows the depth profiles of the saturation index (*omegadel*), the solution rate, and the respiration rate. At the shallowest depths, the saturation index changes rapidly from its supersaturated value at the sediment–water interface, corresponding to seawater values of total dissolved carbon and alkalinity, to undersaturation in the top layer of sediment. Corresponding to this change in the saturation index is a rapid and unresolved variation in the dissolution rate. Calcium carbonate is precipitating

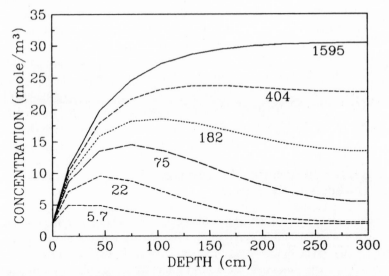

Fig. 8–1. The evolution of the depth profile of total dissolved carbon. The curves are labeled with the time in years since the beginning of the calculation.

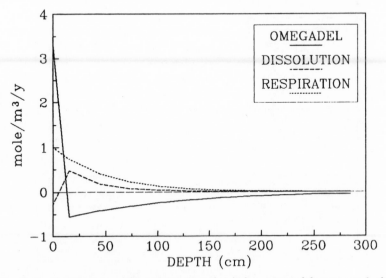

Fig. 8–2. Steady-state profiles of the saturation index, *omegadel = omega*-1, the dissolution rate, and the respiration rate.

at the sediment–water interface and is dissolving in all of the sedimentary layers in this calculation. This transition between precipitation and dissolution needs more careful examination to ensure that matter is conserved by the results. The top layers of the sediment need to be treated with finer depth resolution. But finer depth resolution is not needed deeper in the sediments, where the variations with depth are slow. This calculation would be better with layers of variable thickness. I shall present a modified program in the next section.

8.3 Improving Accuracy by Varying the Thickness of the Layers

The calculation with variable layer thickness is straightforward in principle but requires some care in application. I introduce a number of new arrays that specify layer thicknesses, separations, and depths, as illustrated in Figure 8–3. The values of these arrays' elements are calculated in subroutine SPECS.

Layer thicknesses are specified in the array *layt,* with the values set in

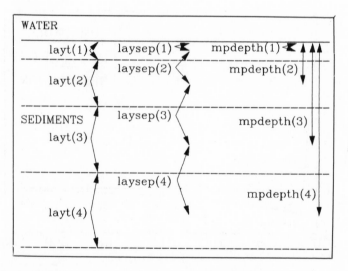

Fig. 8–3. Definitions of layer thickness, layer separation, and depth to the midpoint of the layer.

```
'Program SEDS02 solves diagenesis in carbonate sediments
'Variable layer thicknesses
numlay = 10                  'number of layers
numspec = 2                  'number of interacting species
nrow = numlay * numspec               'the number of equations and unknowns
ncol = nrow + 1
'=====================================================================
DIM sleq(nrow, ncol), unk(nrow), y(nrow), dely(nrow), yp(nrow)
DIM incind(nrow)
'Establish graphics arrays
numplot = 19                 'Number of variables to plot.  Not more than 19
DIM plotz(numplot), plots(numplot), plotl$(numplot), ploty(numplot)
DIM resp(numlay), diss(numlay)    'sedimentary rates
DIM trans(numlay), mpdepth(numlay)
'diffusion coefficient/thickness, depths
DIM layt(numlay), laysep(numlay) 'Thicknesses and separations of layers
'=====================================================================
GOSUB SPECS
GOSUB FILER
xstart = 1                   'time to start
xend = 10000                 'time to stop, in years
x = xstart
delx = 1
mstep = 500
GOSUB CORE
END
'********************************************************************************
EQUATIONS: 'The differential equations that define the simulation
GOSUB OTHER
'Odd numbers are total dissolved carbon, even are calcium
yp(1) = ((sigcs - y(1)) / laysep(1) + (y(3) - y(1)) / laysep(2)) * trans(1)
yp(1) = yp(1) + resp(1) + diss(1)           'continuation
FOR jlay = numspec TO numlay - 1
    jsc = numspec * jlay - 1
    yp(jsc) = (y(jsc - numspec) - y(jsc)) / laysep(jlay)
    yp(jsc) = yp(jsc) + (y(jsc + numspec) - y(jsc)) / laysep(jlay + 1)
    yp(jsc) = yp(jsc) * trans(jlay) + diss(jlay) + resp(jlay)
NEXT jlay
jsc = numspec * numlay - 1
yp(jsc) = (y(jsc - numspec) - y(jsc)) / laysep(numlay)
yp(jsc) = yp(jsc) * trans(numlay) + diss(numlay) + resp(numlay)
'=====================================================================
```

```
'equations for dissolved calcium
yp(2) = ((casw - y(2)) / laysep(1) + (y(4) - y(2)) / laysep(2)) * trans(1)
yp(2) = yp(2) + diss(1)
FOR jlay = numspec TO numlay - 1
    jsc = numspec * jlay
    yp(jsc) = (y(jsc - numspec) - y(jsc)) / laysep(jlay)
    yp(jsc) = yp(jsc) + (y(jsc + numspec) - y(jsc)) / laysep(jlay + 1)
    yp(jsc) = yp(jsc) * trans(jlay) + diss(jlay)
NEXT jlay
jsc = numspec * numlay
yp(jsc) = (y(jsc - numspec) - y(jsc)) / laysep(numlay)
yp(jsc) = yp(jsc) * trans(numlay) + diss(numlay)
RETURN
'*****************************************************************************
SPECS: 'Subroutine to read in the specifications of the problem
secpy = 365.25 * 3600! * 24!    'seconds per year.
difsed = .000002              'cm^2/sec. Broecker and Peng, p. 55
difsed = difsed * secpy       'cm^2/y
prsty = .5                    'porosity
transcon = difsed / prsty          'effective diffusion coefficient
respdec = 50               'respiration decay depth in cm
resp0 = .000001            'respiration rate in mole/cm^3/y
resp0 = resp0 * 1000000!'mole/m^3/y
'layer thicknesses in cm
layt(1) = 1: layt(2) = 2: layt(3) = 4: layt(4) = 7: layt(5) = 15
layt(6) = 30: layt(7) = 50: layt(8) = 100: layt(9) = 200: layt(10) = 400
laysep(1) = layt(1) / 2 'midpoint of top layer to water
FOR jlay = 2 TO numlay           'separation of midpoints
    laysep(jlay) = (layt(jlay - 1) + layt(jlay)) / 2
NEXT jlay
laydepth = 0               'depth to midpoint of layer
FOR jlay = 1 TO numlay
    trans(jlay) = transcon / layt(jlay)      'diffusion/thickness
    laydepth = laydepth + laysep(jlay)
    mpdepth(jlay) = laydepth 'depth to layer midpoints, for printer
    'respiration rate
    resp(jlay) = resp0 * EXP(-laydepth / respdec)
NEXT jlay
casw = 10                  'calcium ions in surface water, mole/m^3
csat = .46
'calcite saturation constant in mole^2/m^6, Broecker and Peng, p. 59
disscon = 7      'dissolution constant in DISS(JLAY)
```

```
pcpcon = 1     'carbonate precipitation constant
disfac = .01     'scaling factor in dissolution rate mole/m^3/y
sigcs = 2.01     'mole/m3
alks = 2.2
watemp = 288    'temperature of surface water, deg K
FOR jlay = 1 TO numlay          'initial values equal sea water values
     y(2 * jlay - 1) = sigcs
     y(2 * jlay) = casw
NEXT jlay
FOR jrow = 1 TO nrow
     incind(jrow) = 1
NEXT jrow
RETURN
'*********************************************************************
```
Plus subroutines GAUSS and SWAPPER from Program DGC03
Subroutine STEPPER from Program DGC04
Subroutine·SLOPER from Program DGC08
Subroutine CARBONATE, GRINC, PLTC, FILER, STOPPER, and STARTER
     from Program DGC09
Subroutine CHECKSTEP from Program ISOT01
Subroutine PRINTER, OTHER, GRAFINIT, PLOTTER, and CORE from Program SEDS01

subroutine SPECS. The parameter *laysep* is the separation of the midpoint of one layer from the midpoint of the adjacent layer. *laysep* is important to the calculation of diffusive transport, which depends on the gradient of concentration, calculated as the difference in concentration in adjacent layers divided by the separation of the layers' midpoints. The layer separation appears explicitly in the equations for rates of change, *yp*, in subroutine EQUATIONS. The respiration rate in each layer is calculated as an exponential function of the depth of the layer's midpoint. The depth of the midpoint is contained in the array *mpdepth*. These values also are the depths associated with the calculated concentrations for plotting purposes. Calculated concentrations are plotted at the layers' midpoint depths. Finally, the quantity that is conserved is the total dissolved carbon or calcium in a column extending from the top of a layer to the bottom, but I have written the equations as rates of change of concentrations expressed in amounts per unit volume. Therefore, the expressions for rates of change, *yp*, are divided by the layer thicknesses.

In program SEDS01, the square of layer thickness appeared in the expression for *transcon* in subroutine SPECS. In program SEDS02, one factor of layer thickness is incorporated in the array, *trans,* which is calculated in subroutine SPECS. The second factor of layer thickness, *layt,* which appeared in *transcon* in SEDS01, has been replaced by *laysep,* the layer separation, in program SEDS02. It appears in the expressions for *yp* in subroutine EQUATIONS.

The results of the calculation with variable layer thickness appear in Figure 8–4. I still have just ten sedimentary layers, but with the variable layer thickness I am now able to extend the calculation to a depth of 800 centimeters without degrading the depth resolution at shallow depths. This deeper sedimentary column takes longer to achieve a steady state than does the 300-centimeter sedimentary layer calculated with program SEDS01. The new calculation has therefore been run for 10,000 simulated years. The final steady-state values of total dissolved carbon against depth are not very different, however. Evidently there is too little respiration going on at depths greater than 300 centimeters to cause much dissolution of calcium carbonate and too little precipitation in the topmost layers of the sediments for finer depth resolution to much affect the results.

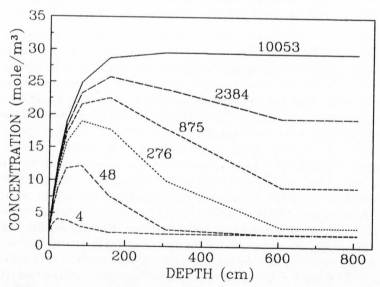

Fig. 8–4. The evolution of the depth profile of total dissolved carbon. The curves are labeled with the time in years since the beginning of the calculation.

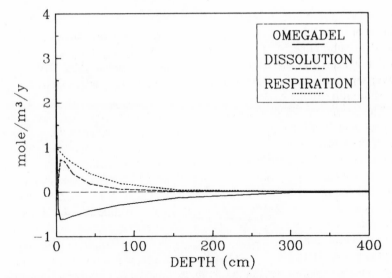

Fig. 8–5. Steady-state profiles of the saturation index, *omegadel* = *omega*-1, the dissolution rate, and the respiration rate to a depth of 400 centimeters.

Figure 8–5 plots the carbonate parameters in the steady state as a function of depth between 0 and 400 centimeters. The figure shows that the saturation index, dissolution rate, and respiration rate all are very close to zero at 400 centimeters. The results for this simulation therefore do not depend on the total thickness of the sedimentary column, provided that this total thickness exceeds 400 centimeters, a limit that depends on the rate at which respiration decreases with increasing depth.

Figure 8–6 shows the parameters of the carbon system as a function of depth just between 0 and 40 centimeters. The new calculation resolves the transition between carbonate precipitation at the shallowest depths and dissolution in the deeper reservoirs. The layer approximation is still apparent in this figure, however, as straight-line segments in the plot. Although it seems unlikely that calculated concentration profiles will be much affected, a further increase in depth resolution at the shallower depths would probably be worthwhile. I shall make this change in the next section, repeating the calculation with fifteen sedimentary layers in place of the ten used so far. First, however, I want to speed up the calculation by avoiding unnecessary computation.

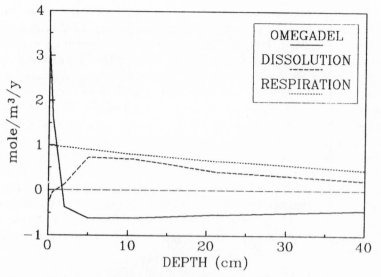

Fig. 8–6. Steady-state profiles of the saturation index, *omegadel* = *omega*-1, the dissolution rate, and the respiration rate to a depth of 40 centimeters.

8.4 Improving Speed by Avoiding Unnecessary Computation

Like the climate system described in Chapter 7, this diagenetic system consists of a chain of identical reservoirs that are coupled only to adjacent reservoirs. Elements of the *sleq* array are nonzero close to the diagonal only. Unnecessary work can be avoided and computational speed increased by limiting the calculation to the nonzero elements. The climate system, however, has only one dependent variable, temperature, to be calculated in each reservoir. The band of nonzero elements in the *sleq* array is only three elements wide, corresponding to the connection between temperatures in the reservoir being calculated and in the two adjacent reservoirs. The diagenetic system here contains two dependent variables, total dissolved carbon and calcium ions, in each reservoir. The species are coupled to one another in each reservoir by carbonate dissolution and its dependence on the saturation state. They also are coupled by diffusion to their own concentrations in adjacent reservoirs. The method of solution that I shall develop in this section can be applied to any number of interacting species in a one-dimensional chain of identical reservoirs.

First, I specify the number of layers by *numlay*, and the number of interacting species by *numspec*. The number of coupled differential equations is *nrow*, which is equal to the product of the number of layers and the number of species. The equations must be arranged systematically, with equations for all species in the first layer followed by equations for all species in the next layer, and so on down the chain of interacting reservoirs. This arrangement was introduced in subroutine EQUATIONS in program SEDS02, in which odd-numbered equations refer to total carbon and even-numbered equations refer to dissolved calcium. An alternative and perhaps more obvious arrangement would be to have the equations for one species at all depths followed by the equations for the next species at all depths. But this arrangement would not leave the desired concentration of nonzero elements close to the diagonal in the *sleq* array. With the equations arranged by species at each depth, the *sleq* array looks like the following, with the nonzero elements extending only a distance of *numspec* = 2 from the diagonal:

$$
\begin{array}{cccccccccccc}
X & X & X & 0 & 0 & 0 & 0 & 0 & . & . & . & X \\
X & X & 0 & X & 0 & 0 & 0 & 0 & . & . & . & X \\
X & 0 & X & X & X & 0 & 0 & 0 & . & . & . & X \\
0 & X & X & X & 0 & X & 0 & 0 & . & . & . & X \\
0 & 0 & X & 0 & X & X & X & 0 & . & . & . & X \\
0 & 0 & 0 & X & X & X & 0 & X & . & . & . & X \\
\end{array}
$$

Other elements of the *sleq* array refer to connections that are not present, either to the given species in more distant reservoirs or to other species in other reservoirs. The elements that describe nonexistent connections are zero and do not need to be calculated. Therefore subroutine SLOPER can be made more efficient by eliminating unnecessary calculations. What is needed is an appropriate alteration of the limits in the loops that carry out the calculations. The modified subroutine SLOPERND appears in program SEDS03.

In the same way, unnecessary work can be avoided by adjusting the limits of the loops in the subroutine that carries out the Gaussian elimination and back substitution. The new subroutine appears in program SEDS03 as GAUSSND. As already noted in Chapter 7, a diagonal system like this one cannot have zero elements on the diagonal, and it will lose its special character if equations are exchanged. Therefore subroutine SWAPPER has been eliminated. The new program with SLOPERND and GAUSSND yields exactly the same results as does program SEDS02 when run with the same layers. Indeed, it must yield identical results because it is performing exactly

```
'Program SEDS03 solves diagenesis in carbonate sediments
'Variable layer thicknesses
'Uses the GAUSSND and SLOPERND solver for a nearly diagonal matrix
numlay = 15              'number of layers
numspec = 2             'number of interacting species
nrow = numlay * numspec          'the number of equations and unknowns
ncol = nrow + 1
'=====================================================================
DIM sleq(nrow, ncol), unk(nrow), y(nrow), dely(nrow), yp(nrow)
DIM incind(nrow)        'Indicates how to test in CHECKSTEP
'Establish graphics arrays
numplot = 19            'Number of variables to plot.  Not more than 19
DIM plotz(numplot), plots(numplot), plotl$(numplot), ploty(numplot)
DIM resp(numlay), diss(numlay)   'sedimentary rates
DIM trans(numlay), mpdepth(numlay) 'diffusion coefficient/thickness, depths
DIM layt(numlay), laysep(numlay)'Thicknesses and separations of layers
'=====================================================================
GOSUB SPECS
GOSUB FILER
xstart = 1              'time to start
xend = 15000            'time to stop, in years
x = xstart
delx = 1
mstep = 500
GOSUB CORE
END
'********************************************************************************
SLOPERND: REM Subroutine to calculate the coefficient matrix, SLEQ
'for a  nearly diagonal matrix
GOSUB EQUATIONS                            'calculate the derivatives
FOR jrow = 1 TO nrow
    sleq(jrow, ncol) = yp(jrow)
NEXT jrow
FOR jcol = 1 TO nrow
    IF y(jcol) = 0 THEN yinc = dlny ELSE yinc = y(jcol) * dlny
    y(jcol) = y(jcol) + yinc
    GOSUB EQUATIONS
    jcll = jcol - numspec: jcul = jcol + numspec     'Limits on JROW
    IF jcll < 1 THEN jcll = 1
    IF jcul > nrow THEN jcul = nrow
    FOR jrow = jcll TO jcul        'differentiate with respect to Y(JCOL)
        sleq(jrow, jcol) = -(yp(jrow) - sleq(jrow, ncol)) / yinc
```

```
        NEXT jrow
        y(jcol) = y(jcol) - yinc      'restore the original value
        sleq(jcol, jcol) = sleq(jcol, jcol) + 1 / delx
            'extra term in diagonal elements
    NEXT jcol
    RETURN
    '***************************************************************************
    GAUSSND: 'Subroutine GAUSSND solves a system of simultaneous linear
    'algebraic equations by Gaussian elimination and back substitution.
    'The number of equations (equal to the number of unknowns) is NROW.
    'The coefficients are in array SLEQ(NROW,NROW+1), where the last column
    'contains the constants on the right hand sides of the equations.
    'The answers are returned in the array UNK(NROW).
    'The array is nearly diagonal, with NUMSPEC nonzero elements
    'off the diagonal
    FOR jrow = 1 TO nrow - 1
        diag = sleq(jrow, jrow)
        jrns = jrow + numspec
        IF jrns > nrow THEN jrns = nrow
        jrp1 = jrow + 1
        FOR jcol = jrp1 TO jrns
            REM divide by coefficient on the diagonal
            sleq(jrow, jcol) = sleq(jrow, jcol) / diag
        NEXT jcol
        sleq(jrow, ncol) = sleq(jrow, ncol) / diag
        sleq(jrow, jrow) = 1
        FOR jr = jrp1 TO jrns
            coeff1 = sleq(jr, jrow)
            REM zeroes below the diagonal
            FOR jcol = jrp1 TO jrns
                sleq(jr, jcol) = sleq(jr, jcol) - sleq(jrow, jcol) * coeff1
            NEXT jcol
            sleq(jr, ncol) = sleq(jr, ncol) - sleq(jrow, ncol) * coeff1
            sleq(jr, jrow) = 0
        NEXT jr
    NEXT jrow
    sleq(nrow, ncol) = sleq(nrow, ncol) / sleq(nrow, nrow)
    sleq(nrow, nrow) = 1
    '=================================================
    'calculate unknowns by back substitution
    unk(nrow) = sleq(nrow, ncol)
    FOR jrow = nrow - 1 TO 1 STEP -1
```

```
    rsum = 0
    jrns = jrow + numspec
    IF jrns > nrow THEN jrns = nrow
    FOR jcol = jrow + 1 TO jrns
        rsum = rsum + unk(jcol) * sleq(jrow, jcol)
    NEXT jcol
    unk(jrow) = sleq(jrow, ncol) - rsum
NEXT jrow
RETURN
'*****************************************************************************
SPECS: 'Subroutine to read in the specifications of the problem
secpy = 365.25 * 3600! * 24!   'seconds per year
difsed = .000002           'cm^2/sec. Broecker and Peng, p. 55
difsed = difsed * secpy    'cm^2/y
prsty = .5                 'porosity
transcon = difsed / prsty          'effective diffusion coefficient
respdec = 50               'respiration decay depth in cm
resp0 = .000001            'respiration rate in mole/cm^3/y
resp0 = resp0 * 1000000! 'mole/m^3/y
'layer thicknesses in cm
layt(1) = 1: layt(2) = 1: layt(3) = 1: layt(4) = 2: layt(5) = 2
layt(6) = 3: layt(7) = 5: layt(8) = 7: layt(9) = 10: layt(10) = 15
layt(11) = 25: layt(12) = 50: layt(13) = 80: layt(14) = 150: layt(15) = 300
laysep(1) = layt(1) / 2 'midpoint of top layer to water
FOR jlay = 2 TO numlay           'separation of midpoints
    laysep(jlay) = (layt(jlay - 1) + layt(jlay)) / 2
NEXT jlay
laydepth = 0               'depth to midpoint of layer
FOR jlay = 1 TO numlay
    trans(jlay) = transcon / layt(jlay)      'diffusion/thickness
    laydepth = laydepth + laysep(jlay)
    mpdepth(jlay) = laydepth 'depth to layer midpoints, for printer
    'respiration rate
    resp(jlay) = resp0 * EXP(-laydepth / respdec)
NEXT jlay
casw = 10                  'calcium ions in surface water, mole/m^3
csat = .46
'calcite saturation constant in mole^2/m^6, Broecker and Peng, p. 59
disscon = 7      'dissolution constant in DISS(JLAY)
pcpcon = 1      'carbonate precipitation constant
disfac = .01    'scaling factor in dissolution rate mole/m^3/y
sigcs = 2.01    'mole/m3
```

```
alks = 2.2
watemp = 288    'temperature of surface water, deg K
FOR jlay = 1 TO numlay          'initial values equal sea water values
    y(2 * jlay - 1) = sigcs
    y(2 * jlay) = casw
NEXT jlay
FOR jrow = 1 TO nrow            'test relative increments in CHECKSTEP
    incind(jrow) = 1
NEXT jrow
RETURN
'*****************************************************************************
Plus subroutine STEPPER from Program DGC04
Subroutine CARBONATE, GRINC, PLTC, FILER, STOPPER, and STARTER
    from Program DGC09
Subroutine CHECKSTEP from Program ISOT01
Subroutine PRINTER, OTHER, GRAFINIT, PLOTTER, and CORE from Program SEDS01
Subroutine EQUATIONS from Program SEDS02
```

the same significant calculations. The only change is the elimination of calculations that do not affect the results. Having tested the program with the layers of SEDS02, I then increase the number of layers to fifteen, to take advantage of the much greater speed of the new program, and I specify new layer thicknesses with improved depth resolution in subroutine SPECS. The modified program appears as SEDS03.

Figure 8–7 shows that the carbonate dissolution parameters vary more smoothly with depth in the new results, and Figure 8–8 shows the approach to the steady state of both total dissolved carbon and calcium ion profiles as functions of depth. Total dissolved carbon is plotted as the solid lines, and calcium as the dashed lines. The profiles are plotted at three different times, measured in years from the start of the calculation. As before, concentrations at first increase most rapidly at the shallow depths, where the respiration rate is greatest. The final concentrations are independent of depth below about 400 centimeters, indicating that there are negligible sources or sinks of dissolved constituents at the greater depths. In this system, the calcium concentration is not quite twice as large at great depths as in the overlying seawater, but the total dissolved carbon concentration increases by a factor of more than 10. Extra calcium is provided only by the dissolution of calcium carbonate, but extra carbon is provided by respiration as well as by dissolution.

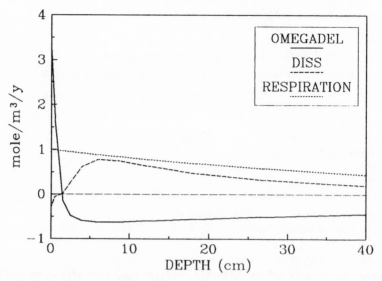

Fig. 8–7. Steady-state profiles of the saturation index, *omegadel* = *omega*-1, the dissolution rate, and the respiration rate to a depth of 40 centimeters. This calculation uses a finer depth resolution.

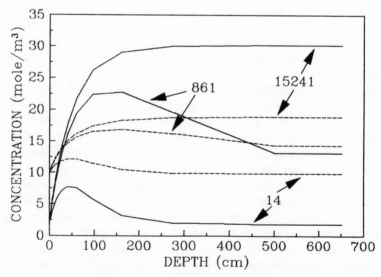

Fig. 8–8. The evolution of the depth profiles of total dissolved carbon (solid lines) and alkalinity (dashed lines). The curves are labeled with the time in years since the beginning of the calculation.

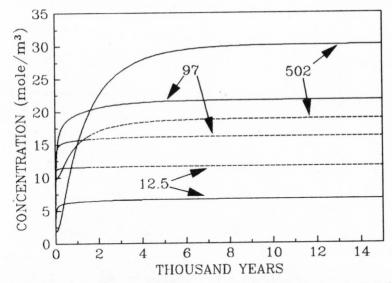

Fig. 8–9. The evolution of total dissolved carbon (solid lines) and alkalinity (dashed lines) at various depths, specified in centimeters.

That the system has indeed reached a steady state can be confirmed by looking at the variation of concentrations at various depths as functions of time, plotted in Figure 8–9. The numbers labeling the curves in this figure are the depths. Total carbon is plotted as solid lines, and dissolved calcium as dashed lines. Concentrations at shallow depths evolve rapidly to nearly their final values. The evolution is slower at the deeper levels, but even at a depth of 500 centimeters there is little change in the concentrations after 10,000 years of simulation time.

Figure 8–10 shows the first 200 years of evolution of the concentrations at the same depths as plotted in Figure 8–9. The concentrations of both total carbon and calcium at a 500-centimeter depth decrease at first and then increase. This decrease occurs because I used starting values equal to seawater values. The waters were initially supersaturated and started out by precipitating calcium carbonate. This initial precipitation was overwhelmed at the shallower depths by the rapid addition of carbon as a result of respiration.

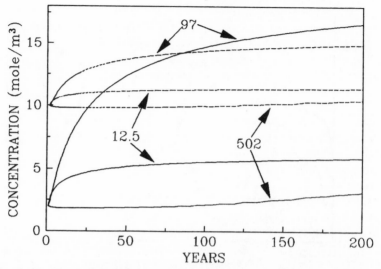

Fig. 8–10. The first 200 years of the evolution of total dissolved carbon (solid lines) and alkalinity (dashed lines) at various depths, specified in centimeters.

8.5 Carbon Isotopes in Carbonate Sediments

Now I shall show how the nearly diagonal system can easily be modified to incorporate additional interacting species. In this illustration I shall add the calculation of the stable carbon isotope ratio specified by $\delta^{13}C$. All of the parameters that affect the concentrations of carbon and calcium are left as in program SEDS03, so that the concentrations remain those that were plotted in Section 8.4. I shall not repeat the plots of the concentrations but present just the results for the isotope ratio.

In program SEDS04 I increase the number of species, *numspec,* to 3 and add the equations for the isotope ratio to subroutine EQUATIONS. Because the equations must be arranged by species in each reservoir, this necessitates some changes in the array indices. Total carbon is now represented by every third equation beginning with equation 1; calcium is every third equation beginning with equation 2; and $\delta^{13}C$ is every third equation beginning with equation 3. The equations for isotope ratio are closely related to the equations for total carbon, but extra terms arise because the isotopic delta is a ratio and is not directly conserved. The whole equation for the rate of change of an isotope ratio, *yp(3),* for example, is divided by the concentration of total dissolved carbon, *y(1),* and there is a term equal

```
'Program SEDS04 solves diagenesis in carbonate sediments
'Variable layer thicknesses
'Uses the GAUSSND and SLOPERND solver for a nearly diagonal matrix
'Includes stable carbon isotopes
numlay = 15              'number of layers
numspec = 3              'number of interacting species
nrow = numlay * numspec             'the number of equations and unknowns
ncol = nrow + 1
'====================================================================
DIM sleq(nrow, ncol), unk(nrow), y(nrow), dely(nrow), yp(nrow)
DIM incind(nrow)         'Indicates how to test in CHECKSTEP
'Establish graphics arrays
numplot = 19             'Number of variables to plot. Not more than 19
DIM plotz(numplot), plots(numplot), plotl$(numplot), ploty(numplot)
DIM resp(numlay), diss(numlay)   'sedimentary rates
DIM trans(numlay), mpdepth(numlay) 'diffusion coefficient/thickness, depths
DIM layt(numlay), laysep(numlay)'Thicknesses and separations of layers
'====================================================================
GOSUB SPECS
GOSUB FILER
'Open a second file for isotope results
OPEN "RESISO.PRN" FOR OUTPUT AS #4
xstart = 1               'time to start, from file
xend = 15000             'time to stop, in years
x = xstart
delx = 1
mstep = 500
GOSUB CORE
END
'********************************************************************
PRINTER:   'Subroutine writes a file for subsequent plotting
GOSUB OTHER                      'print current values
PRINT #1, USING "###.###"; sigcs;
FOR jrow = 1 TO numlay       'Odd numbers are total carbon
    PRINT #1, USING "###.###"; y(3 * jrow - 2);
NEXT jrow
PRINT #1, USING "###.###"; casw;
FOR jrow = 1 TO numlay       'Even numbers are calcium
    PRINT #1, USING "###.###"; y(3 * jrow - 1);
NEXT jrow
PRINT #1, x
'Extra statements to print isotope results
```

```
PRINT #4, USING '###.###'; delcw;
FOR jrow = 1 TO numlay          'isotopes in a different file
    PRINT #4, USING '###.###'; y(3 * jrow);
NEXT jrow
PRINT #4, x
GOSUB PLOTTER
RETURN
'*****************************************************************************
EQUATIONS: 'The differential equations that define the simulation
GOSUB OTHER
'equations for total dissolved carbon
yp(1) = ((sigcs - y(1)) / laysep(1) + (y(4) - y(1)) / laysep(2)) * trans(1)
yp(1) = yp(1) + resp(1) + diss(1)          'continuation
FOR jlay = 2 TO numlay - 1
    jsc = numspec * jlay - 2
    yp(jsc) = (y(jsc - numspec) - y(jsc)) / laysep(jlay)
    yp(jsc) = yp(jsc) + (y(jsc + numspec) - y(jsc)) / laysep(jlay + 1)
    yp(jsc) = yp(jsc) * trans(jlay) + diss(jlay) + resp(jlay)
NEXT jlay
jsc = numspec * numlay - 2
yp(jsc) = (y(jsc - numspec) - y(jsc)) / laysep(numlay)
yp(jsc) = yp(jsc) * trans(numlay) + diss(numlay) + resp(numlay)
'=========================================================
'equations for dissolved calcium
yp(2) = ((casw - y(2)) / laysep(1) + (y(5) - y(2)) / laysep(2)) * trans(1)
yp(2) = yp(2) + diss(1)
FOR jlay = 2 TO numlay - 1
    jsc = numspec * jlay - 1
    yp(jsc) = (y(jsc - numspec) - y(jsc)) / laysep(jlay)
    yp(jsc) = yp(jsc) + (y(jsc + numspec) - y(jsc)) / laysep(jlay + 1)
    yp(jsc) = yp(jsc) * trans(jlay) + diss(jlay)
NEXT jlay
jsc = numspec * numlay - 1
yp(jsc) = (y(jsc - numspec) - y(jsc)) / laysep(numlay)
yp(jsc) = yp(jsc) * trans(numlay) + diss(numlay)
'=========================================================
'equations for delta 13 C
yp(3) = (sigcs * delcw - y(1) * y(3)) / laysep(1)
yp(3) = (yp(3) + (y(4) * y(6) - y(1) * y(3)) / laysep(2)) * trans(1)
'dissolution adds carbon at DELCARB.  Precipitation removes carbon at Y(3)
IF diss(1) > 0 THEN dcd = delcarb ELSE dcd = y(3)
yp(3) = yp(3) + resp(1) * delcorg + diss(1) * dcd
```

```
yp(3) = (yp(3) - y(3) * yp(1)) / y(1)
FOR jlay = 2 TO numlay - 1
    jsc = numspec * jlay
    yp(jsc) = (y(jsc - 5) * y(jsc - 3) - y(jsc - 2) * y(jsc))
    yp(jsc) = yp(jsc) / laysep(jlay)
    yp(jsc) = yp(jsc) + (y(jsp + 1) * y(jsc + 3) - y(jsc - 2) * y(jsc))
    yp(jsc) = yp(jsc) / laysep(jlay + 1)
    yp(jsc) = yp(jsc) * trans(jlay) + resp(jlay) * delcorg
    IF diss(jlay) > 0 THEN dcd = delcarb ELSE dcd = y(jsc)
    yp(jsc) = yp(jsc) + diss(jlay) * dcd
    yp(jsc) = (yp(jsc) - y(jsc) * yp(jsc - 2)) / y(jsc - 2)
NEXT jlay
jsc = numspec * numlay
yp(jsc) = (y(jsc - 5) * y(jsc - 3) - y(jsc - 2) * y(jsc)) / laysep(numlay)
yp(jsc) = yp(jsc) * trans(numlay) + resp(numlay) * delcorg
IF diss(numlay) > 0 THEN dcd = delcarb ELSE dcd = y(jsc)
yp(jsc) = yp(jsc) + diss(numlay) * dcd
yp(jsc) = (yp(jsc) - y(jsc) * yp(jsc - 2)) / y(jsc - 2)
RETURN
'***************************************************************************
OTHER: 'Subroutine evaluates quantities that change with time but that
'are not dependent variables
FOR jlay = 1 TO numlay
    'Carbonate equilibria
    sigc = y(3 * jlay - 2): alk = alks + 2 * (y(3 * jlay - 1) - casw)
    GOSUB CARBONATE
    omegadel = co3 * y(3 * jlay - 1) / csat - 1
    'carbonate dissolution rate
    diss(jlay) = (EXP(-disscon * omegadel) - EXP(pcpcon * omegadel))
    diss(jlay) = diss(jlay) * disfac
NEXT jlay
RETURN
'***************************************************************************
SPECS: 'Subroutine to read in the specifications of the problem
secpy = 365.25 * 3600! * 24!    'seconds per year
difsed = .000002          'cm^2/sec.  Broecker and Peng, p. 55
difsed = difsed * secpy    'cm^2/y
prsty = .5                'porosity
transcon = difsed / prsty        'effective diffusion coefficient
respdec = 50             'respiration decay depth in cm
resp0 = .000001        'respiration rate in mole/cm^3/y
resp0 = resp0 * 1000000!'mole/m^3/y
```

```
'layer thicknesses in cm
layt(1) = 1: layt(2) = 1: layt(3) = 1: layt(4) = 2: layt(5) = 2
layt(6) = 3: layt(7) = 5: layt(8) = 7: layt(9) = 10: layt(10) = 15
layt(11) = 25: layt(12) = 50: layt(13) = 80: layt(14) = 150: layt(15) = 300
laysep(1) = layt(1) / 2 'midpoint of top layer to water
FOR jlay = 2 TO numlay           'separation of midpoints
    laysep(jlay) = (layt(jlay - 1) + layt(jlay)) / 2
NEXT jlay
laydepth = 0              'depth to midpoint of layer
FOR jlay = 1 TO numlay
    trans(jlay) = transcon / layt(jlay)        'diffusion/thickness
    laydepth = laydepth + laysep(jlay)
    mpdepth(jlay) = laydepth 'depth to layer midpoints, for printer
    'respiration rate
    resp(jlay) = resp0 * EXP(-laydepth / respdec)
NEXT jlay
casw = 10                'calcium ions in surface water, mole/m^3
csat = .46
'calcite saturation constant in mole^2/m^6, Broecker and Peng, p. 59
disscon = 7      'dissolution constant in DISS(JLAY)
pcpcon = 1      'carbonate precipitation constant
disfac = .01     'scaling factor in dissolution rate mole/m^3/y
sigcs = 2.01      'mole/m3
alks = 2.2
watemp = 288     'temperature of surface water, deg K
delcw = 2         'delta 13 C is surface water, per mil
delcorg = -10    'delta 13 C of organic carbon
delcarb = delcw 'delta 13 C of carbonate sediments
FOR jlay = 1 TO numlay             'initial values equal sea water values
    y(3 * jlay - 2) = sigcs
    y(3 * jlay - 1) = casw
    y(3 * jlay) = delcw
NEXT jlay
FOR jrow = 1 TO numlay            'test relative increments in CHECKSTEP
    incind(numspec * jrow - 2) = 1          'for SIGC and ALK only
    incind(numspec * jrow - 1) = 1
    incind(numspec * jrow) = .5    'test absolute increments for isotopes
NEXT jrow
RETURN
'*****************************************************************************
Plus subroutine STEPPER from Program DGC04
Subroutine CARBONATE, GRINC, PLTC, FILER, STOPPER, and STARTER
```

from Program DGC09
Subroutine CHECKSTEP from Program ISOT01
Subroutine GRAFINIT, PLOTTER, and CORE from Program SEDS01
Subroutine SLOPERND and GAUSSND from Program SEDS03

to the product of the isotope ratio and the rate of change of carbon concentration, $-y(3) * yp(1)$ in the equation for $yp(3)$, for example. These extra terms were discussed in detail in Chapter 6.

I imagine that respiration adds total dissolved carbon at an isotope ratio of *delcorg*, specified by subroutine SPECS. I further assume that dissolution adds total carbon at an isotope ratio *delcarb*, also specified in subroutine SPECS. The precipitation of calcium carbonate removes total carbon at an isotope ratio equal to the isotope ratio of the pore water. There is no fractionation associated with either dissolution or precipitation in this system. Because the isotopic value associated with precipitation is different from that associated with dissolution, I have to test the sign of *diss* before adding the dissolution term to the equation for the rate of change of the isotope ratio in subroutine EQUATIONS. This test is made in the IF statements in this subroutine.

Because isotope ratios are likely to go through zero, it is not satisfactory to test relative increments of the isotope ratios when determining step size in subroutine CHECKSTEP. Instead, I test on the absolute increment in isotope ratio, requiring that this not exceed 0.5 per mil. Control of the test is exercised by array *incind*, for which the values are set at the end of subroutine SPECS, a procedure discussed in more detail in Chapter 6. For convenience, I modified subroutine PRINTER to write the isotopic results in a separate file for subsequent printing. I put the results into a spreadsheet for plotting, but my software will not import a line of more than 240 characters. So if I try to write total dissolved carbon concentrations, calcium concentrations, and isotope ratios in fifteen reservoirs, all on one line, the line will be too long. I can avoid this difficulty by writing the results to separate files and importing them separately to adjacent areas of the spreadsheet, a contrivance that might not be necessary with different graphic software.

The evolution of the isotope ratio at various depths is shown in Figure 8–11 for the first 200 years of the calculation. The shallowest depths depart little from the seawater value because they are diffusively coupled to open water. Respiration at somewhat greater depths drives the isotope ratio

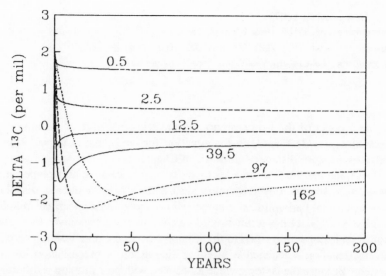

Fig. 8–11. The evolution of carbon isotopes at various depths, specified in centimeters.

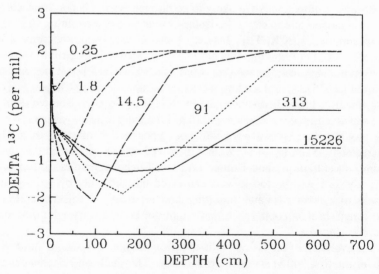

Fig. 8–12. Profiles of carbon isotope ratio at various times, specified in years.

sharply negative at first, but it recovers to more positive values as the dissolution of calcium carbonate releases total carbon with the seawater value.

The evolution of the profiles of the isotope ratio is shown in Figure 8–12, which plots the profiles at various times in the calculation. Early in the calculation, isotope ratios at shallow depths have been driven more negative by the release of isotopically light respiration carbon, but little change has occurred at greater depths. As the evolution proceeds, the ratios at shallow depths become more positive as the result of the dissolution and diffusion of heavier carbon from both above and below. In the final steady state, after some 15,000 years, the isotope ratio is nearly constant at about −0.6 per mil at depths below 100 centimeters, rising rapidly to the seawater value, +2 per mil in the top 100 centimeters. The final values reflect a balance between the release of isotopically light carbon by respiration and the release of isotopically heavy carbon by dissolution, with the additional influence of the diffusion of isotopically heavy seawater carbon.

8.6 Summary

The main goal of this chapter was to demonstrate how to deal effectively with interacting species in a one-dimensional chain of identical reservoirs. The application was to the influence of the oxidation of organic matter on carbonate sediments. I began with a direct application of previously developed computational methods to a system of ten sedimentary layers of equal thickness containing dissolved carbon and calcium ions affected by organic matter respiration and the dissolution and precipitation of calcium carbonate. I then explained how to modify the system for layers of variable thickness, making it possible to improve the depth resolution near the sediment–water interface and to use thicker layers at greater depths, where change with depth is small.

Like the climate system of Chapter 7, this system yields nonzero elements of the *sleq* array only close to the diagonal. Much computation can be eliminated by modifying the solution subroutines, SLOPER and GAUSS. I presented the modified subroutines SLOPERND and GAUSSND, which differ from the equivalent routines of Chapter 7 in that they can accommodate an arbitrary number of interacting species. To illustrate how the computational method can be applied to more species, I added to the system a calculation of the stable carbon isotope ratio, solving finally for the three species, total dissolved carbon, calcium ion concentration, and $\delta^{13}C$, in fifteen sedimentary layers.

Chapter 8 presented the last of the computational approaches that I find widely useful in the numerical simulation of environmental properties. The routines of Chapter 8 can be applied to systems of several interacting species in a one-dimensional chain of identical reservoirs, whereas the routines of Chapter 7 are a somewhat more efficient approach to that chain of identical reservoirs that can be used when there is only one species to be considered. Chapter 7 also presented subroutines applicable to a generally useful but simple climate model, an energy balance climate model with seasonal change in temperature. Chapter 6 described the peculiar features of equations for changes in isotope ratios that arise because isotope ratios are ratios and not conserved quantities. Calculations of isotope ratios can be based directly on calculations of concentration, with essentially the same sources and sinks, provided that extra terms are included in the equations for rates of change of isotope ratios. These extra terms were derived in Chapter 6.

Chapter 5 dealt with the particular complexities of carbonate equilibria in seawater and explained a routine that can be used to solve for dissolved carbon species. A number of useful housekeeping routines also were introduced in this chapter, including routines to coordinate file management and to display results on the screen in graphical form during a calculation. The earlier chapters detailed the numerical solution of a coupled system of ordinary differential equations, beginning with linear systems and ending with a very general approach that can be applied without algebraic manipulation to a wide range of problems concerned with the evolution of physical and chemical properties of the environment.

The applications I presented in this book were chosen to illustrate simple computational methods rather than to constitute serious research simulations of natural systems. The advantage of the simple approach to quantitative simulation of environmental properties is that it is easy to modify a program to test different ideas and different formulations. I have tried to show how easily these modifications can be made, but more numerical experimentation and exploration are needed for all of the systems that I discussed. Such work could well start by building on and adapting the applications programs in this book. For example, my programs might provide reasonable points of departure for the further study of diagenesis in carbonate sediments, seasonal change of surface temperature, the evolution of the carbon system, including isotopes, in response to fossil fuel combustion or geological change, and the ventilation of the deep ocean.

References

Berger, A. L. 1978. Long-term variations of daily insolation and quaternary climate changes. *J. Atmospheric Sci.* **35**, 2362–67.

Broecker, W. S., and T.-H. Peng. 1982. *Tracers in the Sea.* Palisades, N.Y.: Lamont–Doherty Geological Observatory.

Butler, J. N. 1982. *Carbon Dioxide Equilibria and Their Applications.* Reading, Mass.: Addison–Wesley.

Fairbridge, R. W. 1967. Temperature in the atmosphere. In *The Encyclopedia of Atmospheric Sciences and Astrogeology.* Ed. R. W. Fairbridge, New York: Reinhold, pp. 982–90.

Keeling, C. D. 1973. Industrial production of carbon dioxide from fossil fuels and limestone. *Tellus* **25**, 174–98.

Kuhn, W. R., J. C. G. Walker, and H. G. Marshall. 1989. The effect on Earth's surface temperature from variations in rotation rate, continent formation, solar luminosity, and carbon dioxide. *J. Geophys. Res.* **94**, 11129–36.

North, G. R., J. G. Mengel, and D. A. Short. 1983. Simple energy balance model resolving the seasons and the continents: Application to the astronomical theory of the ice ages. *J. Geophys. Res.* **88**, 6576–86.

Parrish, J. T. 1985. Latitudinal distribution of land and shelf and absorbed solar radiation during the Phanerozoic. *Open-File Report 85–31.* Washington, D.C.: U.S. Department of the Interior, Geological Survey.

Rotty, R. M., and G. Marland. 1986. Fossil fuel combustion: Recent amounts, patterns, and trends of CO2. In *The Changing Carbon Cycle, a Global Analysis.* Ed. J. R. Trabalka and D. E. Reichle, New York: Springer-Verlag, pp. 474–90.

Sellers, W. D. 1965. *Physical Climatology.* Chicago: University of Chicago Press.

Siegenthaler, V., and K. O. Münnich. 1981. $^{13}C/^{12}C$ fractionation during CO_2 transfer from air to sea. In *Carbon Cycle Modelling.* Ed. B. Bolin, New York: John Wiley and Sons, pp. 249–57.

Thompson, S. L., and E. J. Barron. 1981. Comparison of Cretaceous and present

Earth albedos: Implications for the causes of paleoclimates. *J. Geol.* **89,** 143–67.

Thompson, S. L., and S. H. Schneider. 1979. A seasonal zonal energy balance climate model with an interactive lower layer. *J. Geophys. Res.* **84,** 2401–14.

Walker, J. C. G. 1977. *Evolution of the Atmosphere.* New York: Macmillan.

List of Programs

Index

Entries in italic lowercase letters are program variables. Entries in roman capital letters are programs, files, subroutines, or acronyms.

185